Ownership and Control of Oil

Ownership and Control of Oil examines government decisions about how much control to exert over the petroleum industry, focusing on the role of National Oil Companies in the production of crude oil since the nationalizations in the 1970s.

- What are the motives for which some producing states opt for less and *not* more control of their oil production sector?
- When can International Oil Companies enter the upstream industry of producing states and under what conditions?

The diversity of policy choices across producers provides the stage for this investigation: different theoretical explanations are confronted with empirical evidence, with the aim of finally proposing an interdisciplinary framework of analysis to explain who controls oil production around the world.

This book is intended for both specialists and general readers who have an interest in the issue of government control of the petroleum industry. Due to its multidisciplinary approach, the book is aimed at a large academic public composed of scholars of political science, international political economy, comparative politics, and Middle East area studies. Moreover, this book should be relevant to international consultants, industry professionals and decision-makers in countries assessing their experience with existing control structures as well as the many countries in the process of joining the "petroleum club" of oil producing nations.

Bianca Sarbu is a post-doctoral researcher at the Center for Security Studies, ETH Zurich. Her research spans the field of Energy Studies with a particular focus on the oil industry, political risks, National Oil Companies and International Oil Companies, as well as the political economy of oil producing countries. For the 2011/2012 academic year, Bianca was a Fulbright Visiting Scholar at the Elliott School of International Affairs, George Washington University, Washington DC, USA.

Ownership and Control of Oil

Explaining policy choices across producing countries

Bianca Sarbu

Routledge
Taylor & Francis Group

LONDON AND NEW YORK

First edition published 2014
by Routledge
2 Park Square, Milton Park, Abingdon, Oxon OX14 4RN

Simultaneously published in the USA and Canada
by Routledge
711 Third Avenue, New York, NY 10017

Routledge is an imprint of the Taylor & Francis Group, an informa business

British Library Cataloguing in Publication Data
A catalogue record for this book is available from the British Library

Library of Congress Cataloging in Publication Data
Sarbu, Bianca.
  Ownership and control of oil : explaining policy choices across
  producing countries / Bianca Sarbu.
     pages cm
  Includes bibliographical references and index.
  ISBN 978-0-415-72599-6 (hb : alk. paper)—ISBN 978-1-315-77809-9
  (ebook : alk. paper)   1. Petroleum industry and trade—Government
  policy.   2. Petroleum industry and trade—Political aspects.   I. Title.
  HD9560.6.S37 2014
  338.2'7282—dc23   2013042670

ISBN: 978-0-415-72599-6 (hbk)
ISBN: 978-1-315-77809-9 (ebk)

Typeset in Sabon
by RefineCatch Limited, Bungay, Suffolk

MIX
Paper from
responsible sources
FSC
www.fsc.org   FSC® C013604

Printed and bound by CPI Group (UK) Ltd, Croydon, CR0 4YY

To my parents and my Manuel

1. The right of peoples and nations to permanent sovereignty over their natural wealth and resources must be exercised in the interest of their national development and of the well-being of the people of the State concerned.
2. The exploration, development and disposition of such resources, as well as the import of the foreign capital required for these purposes, should be in conformity with the rules and conditions which the peoples and nations freely consider to be necessary or desirable with regard to the authorization, restriction or prohibition of such activities.

(United Nations General Assembly Resolution 1803 (XVII)
of December 14 1962,
"Permanent Sovereignty over Natural Resources")

Oil is the most glaring exception to the rules of trade and investment underpinning globalization. (. . .) Barriers to investment are preserved and protected as governments freely discriminate against foreign investment or totally ban it, providing their companies unfettered, uncompetitive monopolies in the exploration, production and sale of oil.

(Edward Morse, SAISPHERE 2005)

Contents

Figures and tables

Figures

Tables

Preface and acknowledgements

This book examines the diverse policies pursued by producing countries in their oil production sector, also known as the oil upstream, since the nationalizations in the 1970s. More specifically, it explores the motivations behind government decisions about how much control to exert over the petroleum industry, focusing on the role of National Oil Companies (NOCs) in the production of crude oil.

The idea of the book concretized in time and took shape in a doctoral research project at the ETH Zurich in Switzerland. Following myriad discussions with fascinating people I had the chance to meet at the UN headquarters during my time as a Youth Delegate of Romania in New York, I grew more and more interested in energy-related topics and resource curse. Curiosity-driven, after my return to the ETH Zurich, I immersed myself in endless readings and debates until I faced the puzzle. In short, a number of oil-rich countries around the world like Kuwait, Saudi Arabia, Iran, Mexico (since the 1990s until the mid-2000s), Uzbekistan and Turkmenistan have kept their oil upstream industry largely under state control. If present, foreign oil companies have operated without decision-making rights over investments, oil prices or profits. In contrast, other oil producing countries like Abu Dhabi in the United Arab Emirates, Oman, Colombia (since the 1990s), Kazakhstan and Azerbaijan have attracted foreign direct investments into their upstream and have made their NOCs share production and profit with International Oil Companies (IOCs). The question that sprang to my mind was "What triggers this variation?" or in a more scholarly form, "Why have oil producing countries pursued diverse policies in their oil upstream sector?". To my surprise, the existing theoretical and empirical scholarship turned out NOT to have a systematic and comprehensive answer to this, which provided the stage for my doctoral investigation and eventually this book project.

Over the past years of research and writing, I have grown grateful to a lot of people for their intellectual and/or moral support. In this sense, I would first like to express my most sincere appreciation and thanks to Professor Robert Weiner from George Washington University in Washington DC for his intellectual guidance, knowledgeable advice, unconditional support, and not least, collegial attitude during my Fulbright Fellowship as well as afterwards as external supervisor on my doctoral committee. For encouraging me to advance my academic degrees, pursue my research interests, find an empirical puzzle, and also, for upholding the trust in my abilities and determination to bring this book project to an end, I am fully indebted to Professor Andreas Wenger from the ETH Zurich.

Furthermore, I am most grateful to Fareed Mohamedi from Statoil and Rodica Donaldson from EDF Renewable Energy (formerly enXco) for offering me invaluable perspectives from energy consultancy and the industry, as well as for introducing me to the community of energy experts on the Middle East. In addition, I am fully indebted to my interview partners, without whose breadth of knowledge, expertise and insights this book would not have been possible.

I should also like to thank Christian Wolf, Raphael Andre Espinoza and Pauline Jones Luong for sharing datasets with me and for fruitful discussions. Valerie Marcel's constructive feedback at the ISA Conference 2010 in Montreal inspired my approach to primary research. The organizers and participants in the "Gulf Energy Challenges" workshop at the Gulf Research Meeting 2012 in Cambridge, UK were particularly helpful in signaling ways to improve the case studies. I owe special debts to all of them as well as to all the others, colleagues and friends, from different universities and organizations who offered constructive feedback and useful comments over time.

Last but not least, it is my family and many dear friends to whom I wish to express my wholehearted thanks for their understanding and support, and for giving me faith and impetus in dwindling times. I finally dedicate this book to my loving parents and to my husband, Manuel, all three of whom are my most ardent supporters and caring crew.

1 Rethinking the ownership and control of oil

Liberalization has penetrated most sectors of the world economy—yet not the energy sector, which remains an exception to the economic rules of free trade and foreign direct investments. National Oil Companies (NOCs) currently own more than 70 percent of global oil reserves and control over 60 percent of the world's oil production (Thurber 2012, 4). This is particularly striking because until less than five decades ago, oil industries were run under the concession system, in which International Oil Companies (IOCs) controlled the hydrocarbons in petroleum-endowed countries. The last wave of nationalizations occurred in the 1970s, when many oil producing countries divested the international oil majors of their petroleum assets and installed monopolies over "exploration, production and sale" from their country. Ever since, by and large, the oil sector has been in the hands of the state. This makes oil—more than any other form of energy—a blatant outlier in terms of state dominance.

The unfolding of the nationalization process started in 1960 with the creation of the Organization of Petroleum Exporting Countries (OPEC) as a joint effort of producer states to collect a higher share of revenues from the foreign oil companies exploiting their national wealth. Two years later, the United Nations (UN) General Assembly Resolution 1803 (XVII) recognized the right of the state and, implicitly, the people to "permanent sovereignty" over their national resources. This marked a particularly important moment in the history of oil because states hosting oil operations of the major corporations had received little of the proceeds derived from their national wealth. The UN Resolution emboldened the states in question to get a grip on their own resources and exploit them to the benefit of their people.

Notably, the levers introduced by the UN Resolution 1803 (XVII) are twofold. Though not named this way, the sovereignty rights proclaimed in the Resolution can be categorized as ownership rights and control rights. While the state—on behalf of the people—is the sole titular of the former, in the case of the latter—as in the rights to explore, develop and produce the resources—the state can decide to either retain them or grant them all or parts thereof to third parties such as privately-owned foreign oil companies. Consequently, while state ownership has become the default among oil producing countries worldwide, state control over oil resources spans the full register from nil to 100 percent.

A "commanding height" (Yergin and Stanislaw 1998), the petroleum sector is critical to the economy of most producer countries. Control rights regimes—whether state, private or mixed—over oil resources can have a widely different impact on the domestic income and economic development of the producer country (Brunnschweiler and Valente 2011). In the oil sector and more generally, mineral resource sector, state control can be exercised through different policy tools: (1) creation of an NOC[1]; (2) contractual arrangements between the host government and IOCs; and/or (3) fiscal systems.

Of the different possible ways to exercise control, this book focuses on the NOCs as a direct means to the end. Alternative policy tools such as the concessionary and contractual systems including fiscal elements are hands-off exercises of control by the state, which raise the question of effectiveness in ruling the oil sector and taking the fair share of oil revenues. History has taught us that NOCs have been preferred by oil producing countries in their endeavor to gain control over national resources and were de facto created for this purpose. As Fadhil J. al-Chalabi, a former Secretary General of OPEC, put this:

> [p]erhaps the most significant development which contributed to the shake-up of the concession system, and which played a great role later on in radically changing the structure of the industry, was the growing trend in the producing countries towards the establishment of national oil companies. These were designed from the beginning to be instruments through which the state could exercise its rights over national resources.
>
> (Quoted in Mahdavi 2011, 11)

NOCs enable government control over the oil sector and its profits in a way that a Ministry of Energy (or equivalent) regulating IOCs does not. State control ensures easier and direct access to oil revenues, which tend to be considerably larger than taxes on IOCs, especially in the short term (Ross 2012, 240). Moreover, IOCs are subject to international accounting standards which impose a certain level of transparency and impede manifest rent-seeking behavior—a practice particularly widespread and favored in authoritarian regimes that many of the oil-rich countries happen to be. Therefore, the NOC becomes a strong (if not the strongest) policy tool for many producer countries in their pursuit of control over petroleum resources.

Across the energy value supply chain, this book focuses on one sector of the oil industry, i.e. the oil exploration and production sector, also known as the oil upstream. *First*, the interest lies in the conventional oil sector because among all forms of energy, "[o]il is unique, not only in the way it is traded and in the way investments are blocked and skewed, but also in the way energy affects other prices, consumer spending, industrial cost structures and ultimately global inflation and economic growth" (Morse 2005). The importance of oil to the world economy is expected to remain paramount over the next decades. Based on "BP Energy Outlook 2030" (2012) and ExxonMobil's "The Outlook for Energy: A View to 2040" (2012), oil and other liquid fuels will continue to be the world's largest energy source for the next two to three decades as they will meet about one-third of the world's energy demand. While accounting for the expansion of both deepwater and unconventional forms of energy, ExxonMobil (2012) predicts that globally, the demand for oil and other liquid fuels will still increase by about 30 percent over the next 30 years, mainly due to commercial transportation.

Second, this book looks into the upstream sector because this is the highest revenue-generating sector across the oil value supply chain. The upstream is considerably more relevant both to oil companies and oil producing countries than the "midstream" (transportation system) or the "downstream" (refining and marketing) sectors, both of which have been historically less lucrative. "Most Middle Eastern producers have low or very low production costs: some $2 for each barrel of crude oil that fetch over $50 on the international market" (Marcel 2006, 5) and $80 to $100 currently. This makes the upstream business not just incredibly profitable (though seriously risk-laden), but also "highly charged politically: it is a central battleground of resource nationalism, which seeks to protect hydrocarbons from foreign hands. By contrast, the downstream business [or similarly, the midstream, *my note*] is not" (ibid., 5). In other words, producer governments are more prone to "discriminate against foreign investment or totally ban it" in the upstream sector (Morse 2005).

Third, this book focuses on the time period since the nationalizations in the 1970s which marks a turning point in the history of the oil industry. For several decades since the beginning of the late nineteenth and early twentieth century, foreign-owned oil companies (IOCs) had dominated the crude oil exploration and production activities all over the world by taking the lion's share out of concessions in regions as different as Latin America, the Middle East or the East Indies. A new twist to the relation between IOCs and host governments came with the renegotiation of the concessional agreements in the 1950s in favor of the latter and the creation of OPEC in 1960. Yet the waning star of the IOCs loomed larger as the 1970s drew closer and the loss of their concessions in the flow of nationalizations became a fact. Most host countries founded their NOCs in this era in two main ways: either from scratch or through the (outright or gradual) divestment of foreign assets. By endowing their NOCs with the former territories of the foreign concessions, host governments paved the way to an oil industry where the IOCs started to take the backseat and the NOCs have increasingly come to the fore. The 1970s thus inaugurated a reverted balance of power between host governments and IOCs.

For all the reasons outlined above, this book examines control structures in the oil upstream sector since the 1970s. In the following, this introductory chapter will first offer some historical notes about the oil industry with a special focus on the evolutionary relation between IOCs and host governments (with respective NOCs). Second, the starting point of the research for this book will be clarified as the empirical puzzle will be mapped out and the research question introduced. Third, the present book will be positioned into the existing literature and the state-of-the-art research. Last but not least, a roadmap will be enclosed to help readers through the next chapters.

1.1 The rising and waning star of IOCs

There are several historical accounts of the oil industry—some of them more extensive and prone to detail than others (Black 2012; Economides and Oligney 2000; Falola and Genova 2005; Marcel and Mitchell 2006; Parra 2004; Yergin 1991, 2009). The historical elements included in this section largely rely on Yergin's comprehensive work and Marcel and Mitchell's short but structured account. The reason for incorporating these selected historical notes here is to provide the general reader with a broad understanding of the changing relations between host governments (with their NOCs) and IOCs. Those readers who are familiar with the topic may like to skip this historical snapshot and move to section 1.2.

The modern history of oil starts in 1859 near Titusville, in Pennsylvania, where the first oil well was dug (Yergin 2009, 270). In the United States, but also Russia (before 1917) and Romania (before 1914), the newly discovered oil properties were quickly bought by a few American and European families—namely, the Rockefellers, Nobels and Rothschilds (Marcel and Mitchell 2006, 16). At the start of the twentieth century, the shiny star of foreign-owned oil companies was on the rise as concessions on very favorable terms were given out and oil was struck in several parts of the world. Governments of developing countries were lacking the capacity to make informed decisions about the discovery and value of their potential national oil resources, which inured to the benefit of IOCs.

The first concession in the Middle East region, which would soon become "the new center of gravity" of oil exploration and production (Yergin 2009, 373), was granted by the Shah of Persia to William Knox D'Arcy in 1901. The Australian-British entrepreneur had no company besides a secretary in charge of his affairs and four years later decided to unite forces with a Scottish oil firm, Burmah Oil Company, in order to be financially able to continue exploring. The discovery of oil in 1908 led to the creation of the Anglo-Persian Oil Company, which would change its name to Anglo-Iranian Oil Company Ltd. in 1935, British Petroleum in 1954 and eventually, BP (Marcel and Mitchell 2006, 16).

The early decades of the twentieth century were molding times for the increase in the use of petroleum and its derivatives. Under Admiral Jack Fisher's influence, Winston Churchill decided to convert the Royal Navy from coal to oil. As Yergin further puts this, "[t]he rapid mechanization of the battlefield in World War I, including the tank and airplane, brought a new mobility to war and made oil an essential strategic commodity" (2009, caption of photos 30 and 31). While until then governments had been interested in the oil resources only marginally, the strategic character that oil was gradually acquiring raised the stakes on the geopolitical scale. The use of oil for kerosene was losing ground due to Thomas Edison's innovation in electricity but new outlets were opening with the development of the automobile industry.

Even before the dawn of World War I, the British government decided to purchase shares in the Anglo-Persian Oil Company[2] in an effort to ensure oil supplies and be independent from prominent foreign oil companies like Royal Dutch/Shell[3] or Standard Oil, i.e. Rockefeller's empire. In 1912, a businessman of Armenian origin, Calouste Gulbenkian succeeded in bringing the interests of the British and German governments under the same umbrella—that is, Turkish Petroleum Company (TPC). The consortium became 50 percent owned by the Anglo-Persian Oil Company, 22.5 percent by the Deutsche Bank, 22.5 percent by Shell, while 5 percent went to Gulbenkian. The aim was to start exploring in Mesopotamia, i.e. present-day Iraq. Quite interestingly, the signatories obliged themselves to the so-called "self-denying cause," which implied that "[n]one would be involved in oil production anywhere in the Ottoman Empire—save jointly 'through the Turkish Petroleum Company'" (ibid., 172). Later known as the Red Line Agreement, this turned out to be "a very significant obligation, one that would haunt many people down through the decades" (ibid., 171). The territories which did not come under this clause were Egypt, Kuwait and "the 'transferred territories' on the Turco-Persian border" (ibid., 172). Thereby, the operational premises for oil exploration and production in the Middle East were set for many years onwards.

As World War I was unfolding, exploration activities had to be postponed. The end of the war brought the defeat of the Ottoman Empire and the share-out of its territories between Britain and France, with a few exceptions—namely, Turkey, Yemen and future Saudi Arabia—which would become independent states. Thus, Britain started to control Iraq, Jordan and Palestine under the League of Nations mandates, and also became "kingmaker in Iran, mandatory power in Iraq and protector of Kuwait and the Trucial states" (Marcel and Mitchell 2006, 17). The British government projected its power in all these territories insofar as the latter could not have any affairs with a foreign entity without prior approval. On these grounds, the oil concessions in these countries came altogether under the direction of the British majority-owned oil company—that is, Anglo-Persian Oil Company. France, in turn, administered Syria and Lebanon.

More and more aware of the oil weapon, the French government supported in 1924 the creation of Companie Française des Pétroles (CFP), later Total. Although the company was fully privately owned, it was backed by the government and represented its interests (Yergin 2009, 174f.).

At the end of the war, as the British, German and French companies were trying to make headway to the Middle East oil, the United States was witnessing a shortfall in fuel oil supplies and the prospect of becoming an oil importer. Washington DC felt that it was losing ground in the Middle East and thus requested "equal access for American capital and business" (ibid., 179) in support of the American companies that were seeking oil supplies abroad.[4] Mesopotamian oil increasingly captured the attention of the British, French and Americans. After the war, the Germans were bought out in the TPC consortium by the French, yet in 1928 the United States questioned the validity of the TPC concession in Iraq. Following intense negotiations, TPC became Iraq Petroleum Company (IPC), in which Royal Dutch/Shell, Anglo-Persian Oil Company, CFP and the Near East Development Company—reuniting American oil companies—owned 23.75 percent each, with the rest of 5 percent interest pertaining to Gulbenkian. The "self-denying clause" remained valid (ibid., 188). The IPC cartel further took over the main concessions in all the territories under direct or indirect British rule—namely, Abu Dhabi, Dubai, Yemen, Oman and Qatar. In brief, "Britain used its political influence to promote large concessions, which favored the concentration of control into the hands of a few foreign companies rather than promoting competition among them" (Marcel and Mitchell 2006, 18).

Outside of the British sphere of influence, inner Arabia allowed for the development of oil in a different manner than in the rest of the Middle East. More explicitly, the Saudi rulers were in the position of choosing whichever company offered the highest pitch to explore and produce oil in the country. This turned out to be Standard Oil of California (Socal), one of the many successor companies following the breakup of Rockefeller's Standard Oil in 1911, which would later be renamed Chevron (Yergin 2009, 273f.). In the late 1930s, Socal combined its forces in Saudi Arabia with an American oil retail company, Texas Oil Company (Texaco) and two Standard Oil spin-offs, i.e. Standard Oil of New Jersey (later Exxon) and Socony-Vacuum (initially, Standard Oil of New York, later renamed Mobil).

The Americans managed to gain more ground in the Middle East given that Britain decided to put aside the British Nationality Clause in Kuwait and Bahrain despite its sphere of influence. Kuwait Oil Company, a joint venture between Gulf Oil Company (an American family company) and Anglo-Persian, negotiated a 75-year concession for an upfront payment of £35,700 to £179,000 (ibid., 280). As for Bahrain, Socal secured the first oil concession and soon thereafter set up a subsidiary, Bahrain Petroleum Company (ibid., 264f.).

By the late 1920s and the early 1930s, IOCs reached the peak of their power and sought to avert the development of an open international oil market where prices could be influenced by increasing production in e.g. Mexico, Venezuela or elsewhere. In this sense, the Achnacarry or the "As-Is" Agreement is emblematic for the endeavor of the large foreign oil companies and their corresponding governments to control the oil industry by setting a fixed or benchmarked oil price and allocating quotas to each other based on their share in 1928[5] (ibid., 251).

The power of the IOCs was not to hold for long though. In the 1930s, resource nationalism was growing and host governments increasingly put pressure on foreign oil companies by setting import quotas and restrictions on foreign exchange, and also by trying to influence prices (ibid.). Tensions were already sizzling under the surface as host governments were not satisfied with their share from the oil income—so it did not take long until concessions were challenged.

The first time that the issue of ownership and control of the oil resources was explicitly addressed in the relations between host governments and oil companies was in the Western Hemisphere in "paragraph 4 of Article 27 of the Mexican constitution of 1917," i.e. "the clause that declared that the underground resources—the 'subsoil', as it was called—belonged not to those who owned the property above, but to the Mexican state" (ibid., 255). Initially, this legal provision did not concretize in any way because investments in the infrastructure of the oil industry were direly needed. However, feelings of adversity against foreign oil companies were growing throughout Latin America and in 1937 the military government of Bolivia suddenly divested Standard Oil's assets on charges of tax fraud. The expropriation was positively received among the people not just in Bolivia but in the entire region. Taxes, royalties as well as the legal status of ownership were large sources of dissatisfaction for the Mexican government too. The general belief among the people was that the American companies were trying to stall the economic growth and curb political stability in the country. On these premises, when the Mexican president announced the expropriation of foreign oil companies in March 1938, "[h]is words were greeted with a six-hour parade through Mexico City" (ibid., 259).

Following this, the Venezuelan government felt emboldened to renegotiate the initial concessions and change the legal framework for the petroleum sector. More explicitly, the hydrocarbons law of 1943 proclaimed the ownership right of the state over the subsurface resources, reduced the concession period to 40 years, raised the government take[6] to one sixth, and imposed a 12 percent corporate tax in all economic sectors. Just four years later, in 1947, the corporate tax was increased more so that the government take would add up to 50 percent of the oil profits (Marcel and Mitchell 2006, 20).

The renegotiation of the concession terms in Iran turned out to be more intricate. In the 1930s, when oil prices were falling and depression was a real menace for the Iranian government, the Pahlavi shah Reza Khan first cancelled the concession altogether before reaching an agreement one year later, in 1933, which encompassed both a cut-down of the initial concession area to only one quarter and an increase in the government take. Given the so-called "new deal in oil" (Yergin 2009, 413)—that is, the amount of government take renegotiated by Venezuela in 1947 and three years later by Saudi Arabia—the Iranian government also pleaded for a 50–50 profit share, yet with no success. This finally led to the nationalization of the Anglo-Iranian concession in 1951 and the creation of the National Iranian Oil Company (NIOC) (Marcel and Mitchell 2006, 20f.).

Notably, these developments spurred renegotiations of the concessional terms in other developing countries all around the world. The star of the IOCs was starting to wane. In 1960, when foreign oil companies reduced the posted crude oil prices, on the basis of which government revenues were calculated, producer countries decided to collaborate. Government representatives of Iraq, Iran, Saudi Arabia, Kuwait and Venezuela came together in Baghdad to form OPEC. Nine other producer countries—namely, Qatar (1961), Indonesia (1962–2009), Libya (1962), United Arab Emirates (1967), Algeria (1969), Nigeria (1971), Ecuador (1973), Angola (2007) and Gabon (1975–1994)—became member states thereafter.[7]

The first decade in the history of OPEC saw an improved standing of host governments in the member countries to the financial detriment of the so-called "Seven Sisters"[8] and other relatively smaller IOCs. However, no radical changes occurred. These would wait until the 1970s when in a concerted action the OPEC member states took control over their national oil industries either through outright nationalization or increased participation[9] (Yergin 2009, 566f.).

The wave of nationalizations set a new milestone for the oil industry. Just four decades later, the *Financial Times* was introducing the "New Seven Sisters" in control of the world's oil reserves and production—all of them NOCs from countries outside the OECD.[10] This book deals with the time period since the 1970s and seeks to understand how producer countries have evolved from this common starting point (i.e. nationalizations) and more importantly, what has made them take different policy paths in their oil upstream sector.

1.2 Empirical puzzle and research question: the picture since the 1970s

Since oil nationalizations in the 1970s, the empirical picture of oil upstream industries around the world—not so much in terms of ownership but more so as concerns control structures[11]—looks very diverse. In the Middle East and North African (MENA) region, a number of oil producing countries like Iran, Kuwait and Saudi Arabia have kept their oil upstream sector largely under state control ever since. When present, foreign oil companies have operated without control rights—under service contracts[12] or equivalent. The actual decision-maker over investments, production, costs and revenues in the oil upstream sector has been the government aided by its NOC—namely, the National Iranian Oil Company (NIOC) in Iran, Kuwait Petroleum Corporation (KPC) in Kuwait and Saudi Aramco in Saudi Arabia. By contrast, Algeria, Oman, Egypt and Abu Dhabi in the United Arab Emirates (UAE) have made their NOCs share production and profit in the upstream with IOCs. Quite uniquely, despite the complete nationalizations in the region, Abu Dhabi has never increased its control over the oil sector to 100 percent but instead bequeathed the IOCs control over approximately 40 percent of production. The 60:40 scheme has been maintained to the present.[13] As for Algeria, Oman and Egypt, they have gradually reopened their oil upstream and relented control rights to private oil companies over the past three decades.

Going beyond the MENA borders, a recent example of increase in state control over oil production is Argentina through the nationalization of the country's largest oil company, Yacimientos Petroliferos Fiscales (YPF) (*The Economist*, April 21, 2012). The Argentinian case of wrenching control from a foreign oil company—in this case, Repsol—is not singular in Latin America. In the past decade, the liberalization measures of the energy sector from the 1990s have been reversed in Venezuela, Bolivia and Ecuador, yet not in Trinidad and Tobago, Brazil and Colombia (Benton 2008, 22). An outlier, Mexico had retained complete state control over its oil upstream sector since 1938 (EIA 2012a) until late 2013.

In the post-Soviet space, the control regimes present in the oil upstream sector also vary greatly. Since 1991, Kazakhstan has managed to attract foreign investments into the oil sector through production-sharing agreements (PSAs). Since 2010 it has done so through joint ventures[14] (EIA 2012b). Similarly, through the implementation of PSAs in 1994, Azerbaijan has engaged IOCs, grouped into Azerbaijan International Operating Company (AIOC), into partnerships with the State Oil Company (SOCAR) in the oil upstream sector (Ciarreta and Nasirov 2012). By comparison, Uzbekistan and Turkmenistan retained the sector under complete state dominance until the mid/late 2000s when they slowly started to release control to private oil companies (EIA 2012c, 2012d). A mixed case, Russia has been oscillating between state and private control—the latter both domestic and foreign. By the end of the 1990s, most of the Russian oil industry was privatized to domestic private oil companies (Weinthal and Jones Luong 2006, 43), which consolidated themselves into a few privately-owned mammoths (EIA 2012e). Despite blatant difficulties, IOCs have been present as well. Just last October, the Russian government increased its control over the oil production sector through the takeover of TNK-BP[15] via Rosneft.[16] Along these lines, the *Financial Times* was writing that the buy-out of both BP and AAR[17] in TNK-BP will "tighten Kremlin's grip on Russia's energy sector" (October 22, 2012).

As noted before, the oil upstream sector is of strategic importance to the economy of oil producing countries. Given the widespread understanding that oil resources need to be produced to the benefit of the people, "the perception that foreign investors could not be trusted to develop resources in the national interest" (Kobrin 1984a, 146) and yet the evidence that some of the oil producing countries have granted control rights to foreign oil operators, the research question guiding this book is: *Why have oil producing countries pursued diverse policies in their oil upstream sector?*

Two conceptual remarks are here in order. First, for a country to be considered an oil producer, it needs to produce crude oil domestically.[18] Second, the control structures in the oil upstream sector are the product of oil upstream sector policies.

Upstream sector policies and thereby, control structures in the oil upstream sector are relevant both for economic and political reasons, as some scholarship on the effects of ownership and control has shown. From an *economic* perspective, private control is considered to be more efficient in the exploitation of the oil resources (Al-Obaidan and Scully 1992; Eller *et al.* 2011; Victor 2007; Wolf 2009). Higher efficiency may imply more revenue for the state. Yet the downside of assigning control to foreign oil companies comes from the repatriation of the residual profits which are often reinvested abroad. "Foreign-based firms have little interest in raising domestic welfare in the host country as this is beyond the scope of their profit-maximization obligation towards shareholders" (Brunnschweiler and Valente 2011, 3). In turn, state control or synonymously, an oil upstream industry operated through an NOC, enables the pursuit of the national interest through the establishment of forward and backward linkages between the oil sector and the rest of the domestic economy (Tordo *et al.* 2011).

From a *political* perspective, there is clear evidence that states are interested in retaining control in strategic sectors, i.e. in "sectors that are essential for economic or physical security such as banking or utilities, and extractive sectors such as mining and petroleum" (Kobrin 1984b, 337). In a study of OECD economies, Bortolotti and Faccio find that "through either direct ownership, or leveraging devices or golden shares, governments maintain control of almost two-thirds of privatized firms" (Bortolotti and Faccio 2009, 2910)—for example, in Eni (the largest oil and gas company in Italy), Enel (Italian electricity company), Alitalia and Deutsche Lufthansa (the largest airline companies in Italy and Germany, respectively). Whether it is political clout, national security considerations, national pride or resource nationalism in the case of extractive sectors (Kobrin 1981, 1984a; Krapels 1993), the motives are not clear.

On these premises of varied empirics as well as in acknowledgement of the political and economic relevance of control in the upstream sector, this book seeks to identify the determinants of different upstream control structures in oil producing states around the world.

1.3 The research relevance

This book tries to make both a theoretical and an empirical contribution to the existing research. *Theoretically*, it explains an economic policy which has been understudied so far—that is, upstream sector policy—and proposes a new focus in International Political Economy—namely, sectoral or industry control. It is an interdisciplinary study which bridges the fields of (International) Political Economy (IPE), Political Science and Regional Studies. Furthermore, it is one of the first systematic scholarly endeavors to analyze the distribution of petroleum exploration and production rights in oil producing countries worldwide.[19]

In a theory-building endeavor, this book puts forward an analytical framework to explain variation in upstream sector policies across oil producing countries worldwide. To this end, it brings together three bodies of literature: the Resource Curse literature; the NOCs literature; and the Nationalization/Expropriation literature. A fourth body of literature is used to strengthen the case studies in the empirical part—that is, Middle East Area Studies. The need for such an analytical framework and the position that this book takes in the existing scholarship are schematically represented in Figure 1.1.

To start with, the *Resource Curse* literature seeks to explain how a country's natural resource wealth shapes its institutions (regime type), economic development, and may lead to political violence (civil war). Ownership and control structures and thereby, the role of policy in the hydrocarbons sector, have been completely neglected by the Resource Curse scholarship.

This is the niche which Pauline Jones Luong and Erika Weinthal's work tries to fill by arguing that "ownership structure is the key intervening variable between mineral wealth and institutional outcomes" (2010, 27). In other words, the curse is not the natural resource endowment but instead, the ownership structure that these mineral-rich countries choose in order to manage their resources. The scholars claim that there is large variation in ownership structure across oil producing countries, which the Resource Curse literature has overlooked because of its (almost) exclusive focus on the same historical period from the late 1960s to the early 1990s.

> It provides a skewed picture of the empirical reality because this is also the time period during which the vast majority of mineral-rich countries did, in fact, exercise state ownership over their mineral resources. Thus, with few exceptions, ownership structure has heretofore been viewed as a constant rather than a variable.
>
> (ibid., 7)

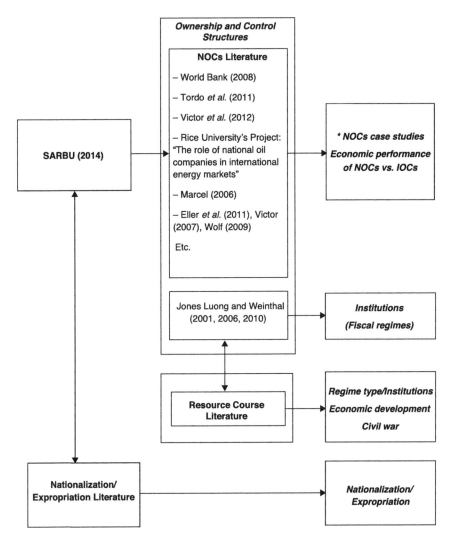

Figure 1.1 Position of this book in relation to the existing literature

While innovative in its idea, Jones Luong and Weinthal's book does not discriminate between state ownership and state control. Along these lines, Mahdavi points out that "Luong and Weinthal's analysis is difficult to confirm given the vague and multidimensional terminology of their coding: in particular, the concept of 'state control' is difficult for the reader to grasp and for the researcher to operationalize" (2011, 10). As a matter of fact, the variation in ownership structures across oil producing countries after the 1990s has been very limited as state ownership has remained prevalent. The actual variation lies in control structures, as observed by other scholars in reference to Jones Luong and Weinthal's work:

> Given this basic assignment of ownership over natural resources to the State [through UN Resolution 1803 (XVII) of 1962, *my note*], the salient question becomes who has the right to exploit these resources, or alternatively: who has access to and control over the resources.
>
> (Brunnschweiler and Valente 2011, 2)

Therefore, state control and thereby, control structures need to be further unpacked and more precisely delimited.

Furthermore, while the Resource Curse literature does not take note of ownership and control structures at all, the *NOCs literature* takes them as given and looks into their effects by comparing the economic performance of NOCs with that of IOCs. Most of this literature is empirically driven and based on case studies (e.g. Rice University's Project on "The Role of National Oil Companies in International Energy Markets"; Marcel 2006; Victor *et al.* 2012), with some exceptions which quantitatively measure performance and efficiency differentials of ownership structures (e.g. Eller *et al.* 2011; Victor 2007; Wolf 2009). The World Bank study (2008) and Tordo *et al.* (2011) are slightly different in their approach to firm performance in the sense that they account for both commercial and non-commercial goals pursued by NOCs—as part of their national mission—by comparison with privately-owned IOCs.

On these grounds, this book takes one step back as compared to all the above-mentioned bodies of literature and explains (ownership and) control structures. The scholarship which comes closest to the focus of this book is the *Nationalization/Expropriation literature*, which examines why countries nationalize. Nationalization[20] is yet different from ownership and control structures. As such, nationalization is just "one option for gaining control over foreign direct investors" (Kobrin 1984b, 329). It is defined as "expropriation or forced divestment of private ownership" by a host government (Kobrin 1980, 65). While nationalization is regarded as a "rare event" and coded as a dichotomous variable in the existing literature, control is the product of an economic policy, which involves a longer time horizon and is a continuous variable along the interval [0, 1] or in percentages [0, 100]. Moreover, while nationalization involves coercion and it is seen as an involuntary divestment, increase in control may be associated with a market-based transaction.[21] In general, nationalizations have a disruptive character for foreign investments in host countries and a negative impact on the government's reputation. By comparison, increase in state control usually functions according to market rules and has no unsettling repercussions.

In the absence of scholarship on control structures, this book draws inspiration from *the Nationalization/Expropriation literature* for the identification of the driving forces of upstream sector policies, the setup of the analytical framework as well as the formulation of hypotheses. However, critical thinking is used to identify potential factors which might work differently in the case of increase in state control than in the case of nationalization. Of all the strands of literature spanned in this work, the Nationalization/Expropriation literature is of primary importance. Additionally, *the NOCs literature* provides insights into the role and relevance of NOCs for oil producing countries, which helps with the conceptualization of upstream sector policy and its operationalization as control structures. Not least, this empirical scholarship offers considerable secondary data for the case studies. *Middle East Area Studies* further contribute to the discussion of the historical background for the case studies and also to the understanding of the regional specificities and their influence on economic policies. Finally, *the Resource Curse literature* helps to map out the context of oil rentier countries like the analyzed cases of Saudi Arabia and Abu Dhabi/the UAE. Its use to this book is yet limited. For a schematic overview of the use of different bodies of literature to this work, please see Table 1.1.

Empirically, the proposed analytical framework is validated in a mixed-methods design which combines advanced statistical methods with two case studies from the Middle East. While the statistical analysis seeks to identify the determinants of control structures in the oil upstream sector globally and thereby, explain the role of NOCs worldwide, the case studies examine how the Middle East context sheds light on these aspects.

The motivation for the regional focus in the case studies is straight-forward: almost 60 percent of the world's proven oil reserves are currently

Table 1.1 Relevance of the existing literature to this book

Body of Literature	Use
Nationalization/ Expropriation	• Identification of the explanatory variables and formulation of hypotheses—with the observation that Nationalization/Expropriation is different from control (Ch. 3.1); • The basis for the analytical framework (Ch. 3.2).
NOCs	• Insights into the role of NOCs and upstream sector policy: the basis for conceptualization and operationalization (Ch. 2); • Secondary data for the case studies (Ch. 5).
Middle Eastern Area Studies	• Historical background for the case studies (Ch. 5.3.1 and 5.4.1); • Perspectives over the Middle Eastern society, economy and political life for the case studies (Ch. 5).
Resource Curse	• Understanding of the wide context: particularly for the case studies but more broadly, for the analytical framework as well.

located in the region, about 30 percent of the world's daily oil is produced in the Middle East[22] (BP 2011) and some of the world's largest unlisted NOCs are Middle Eastern (*Financial Times* Non-Public 150, 2006). It is a matter of magnitude which turns the Middle East region into "the center of gravity of world oil production" (Everette Lee DeGolyer quoted in Yergin 1991, 393) and makes it worth studying.

Due to data limitations, the large-N study in this book covers the timeframe 1987 to 2010. The dataset is compiled from eight different sources and also includes five self-coded variables. For the operationalization of oil upstream sector policy,[23] this work proposes its own measure: *oil upstream industry control* as state versus private control of oil production—in the producer country by year. A doubly-censored Tobit model with a dynamic panel fractional estimator (Elsas and Florysiak 2010, 2012) is then applied for testing.

For the case studies, the analysis goes back to the 1960s–1970s, when the two modern Middle Eastern states (Saudi Arabia and Abu Dhabi/the UAE) were created. This is not a historical book, yet the case studies seek to provide enough historical information for the reader to understand the policies within their actual context. The case studies draw on both secondary literature and primary data collected through expert interviews. In this regard, it should be emphasized that there is very little available information on the inner workings of the oil upstream sector and related policy-making for Saudi Arabia, and even less for Abu Dhabi/the UAE. Based on the significance of these countries in the international energy market and not least, the size of their NOCs, the wealth of primary data shared in the case studies should be of high relevance not just to the Middle East scholars but also to industry professionals and consultants who grapple with energy issues in the region. Some 33 experts from fields as different as the petroleum industry, government, energy consulting, international organizations, think-tanks and academia were interviewed in the course of 2011 and 2012. This primary data was systematized, analyzed and presented in the two case studies which build the second part of the empirical analysis.

Finally, this book should be of relevance both to policy-makers and oil market analysts. To the former, it can provide a better understanding of why and when countries might be protectionist of their oil upstream sector and not least, make them ponder over the implications for the energy security agenda. To the latter, the empirical results pin down the conditions under which foreign investments in the oil upstream are at all possible.

1.4 Roadmap

This book proceeds as follows:

Chapter 2: Upstream sector policy in the oil industry seeks to conceptualize the object of this research: oil upstream sector policy. To this end, the first part maps out the broader context in which upstream sector policies emerge by discussing the institutional set-up for policy-making and thereby, the legal framework given the commercial and/or non-commercial objectives pursued by oil producing countries. The second part theorizes the concept of upstream sector policy by addressing its dimensions and variants, and not least, looks into the tools through which upstream sector policy is "de facto" implemented in oil-rich countries. Due to the meager scholarship on upstream sector policy, such a conceptualization endeavor is imperative before any further research is undertaken.

In search for potential determinants of upstream sector policies, *Chapter 3: Current readings of energy studies and theory development*, reviews the interdisciplinary body of Energy Studies while grouping them according to technical, economic and institutional arguments. In reference to these different explanations, a comprehensive, state-centered analytical framework is then proposed. Starting off with the context, this is defined by three exogenous elements: geological conditions, the technical capabilities of the NOC and the international oil price. The context is meant to create the basis for decision-making and should inform economically efficient policies in the upstream sector. However, the state leader and the executive are primarily interested in political survival. This is why they are likely to make policy choices which are economically inefficient or less efficient as long as these would serve their main goal of remaining in power. Nonetheless, these policy options are strongly constrained by two key domestic forces— i.e. the economic dependence of the country on oil revenues and the limits on the executive power. Finally, the chapter discusses how, and also why, this framework of analysis is applicable to oil producing countries globally.

The proposed analytical framework is further tested in a mixed-method design, which begins with the statistical analysis based on secondary data. Thus, *Chapter 4: A statistical analysis of oil upstream sector policies across the world* introduces the sampling strategy and discusses the process of compiling the dataset. Second, the discussion of data operationalization and sources follows suit in a systematic manner, in reference to other well-established quantitative works. Third, given the newness of the applied method (i.e. Tobit analysis with dynamic panel fractional estimator), its added value and limitations, as compared to an alternative method for the panel in question, are presented in detail. Fourth and finally, the empirical findings are put forward and critically discussed.

Following in the lines of the qualitative research tradition, case studies build the second part of the empirical analysis. In *Chapter 5: Mirroring the cases of Saudi Arabia and Abu Dhabi*, the case selection is first explained with respect to the statistical findings and also, the empirical relevance of the cases in question. The chapter then introduces the primary data collection method, i.e. semi-structured expert interviews, spells out the advantages of this approach to the present research, discusses the structure of the applied questionnaire as well as the professional distribution of the 33 high-level interviewees. The gathered information is systematized, analyzed and presented in the individual case studies on Saudi Arabia and Abu Dhabi. Both start with a brief historical account which contextualizes the oil industry and its growth in the Kingdom and respectively the Emirate, and continue with the discussion of the proposed analytical framework based on primary insights and when available, secondary literature.

Based on a thorough comparison between the statistical results and the insights derived from the two case studies, *Chapter 6: Conclusions*, revises the initially proposed analytical framework by amending and enhancing some of the arguments. The chapter further identifies the contributions as well as the limitations of this book, spells out its relevance for the current political and economic developments in several hotspots across the globe, and last but not least, proposes several research ways forward.

Notes

1 NOCs refer to oil companies under full or partial state ownership. Most countries have or have had an NOC in the past.
2 The interest of the British government in Anglo-Persian Oil Company would be as high as 51 percent (Marcel and Mitchell 2006, 16).
3 Royal Dutch/Shell was created in 1907 through the merger of the Dutch exploration and production company, Royal Dutch Petroleum Company, with the British transport and trading firm, Shell (Yergin 2009, 110).
4 For more on the "Open Door" principle and the repartition of the spheres of influence after World War I, please see Yergin (2009), Ch. 10–14.
5 Due to the increasing production of Russian, Romanian and smaller American companies, the effort of the largest oil companies at the time, which later came to be called the "Seven Sisters" (see endnote 9), did not bear fruit.
6 Government take is the size of the government share from the oil operations in a given country. For a more detailed discussion of the concept, please see Chapter 2.
7 For this, please refer to the OPEC Official Website—Brief History, www.opec. org/opec_web/en/about_us/24.htm, last accessed on September 9, 2013.
8 The "Seven Sisters" is a phrase introduced by Enrico Mattei in the 1950s and is used to describe the major Anglo-Saxon companies that controlled the Middle Eastern crude oil after World War II. The "Seven Sisters" included: Standard Oil of New Jersey (Exxon), Socony-Vacuum (Mobil), Standard Oil of California (Chevron), Texaco, Gulf Oil, Royal Dutch/Shell, and the Anglo-Persian Oil Company (British Petroleum) (Yergin 1991, 503).
9 For an account of the individual cases of outright nationalization and increased participation, please refer to Yergin 2009, Part V.
10 The "New Seven Sisters" are Saudi Aramco (Saudi Arabia), JSC Gazprom (Russia), CNPC (China), NIOC (Iran), PDVSA (Venezuela), Petrobras (Brazil) and Petronas (Malaysia) (Hoyos, in the *Financial Times*, March 11, 2007).
11 For a conceptual apparatus of ownership and control structures and the rationale behind, please see Chapter 2.
12 Under this form of contracts, the IOCs execute clearly defined services. This may include the production of mineral resources, for which they receive a pre-established fee.
13 In other words, Abu Dhabi National Oil Company retains 60 percent in any oilfields operations while the IOCs are granted the remaining 40 percent. It

should also be noted that as of early 2014, the situation in the onshore became slightly unclear. ADNOC took over the onshore operations through ADCO, yet negotiations for new concessions between government and IOCs are currently ongoing and may take until early 2015 (at the latest).

14 Both production-sharing agreements and joint ventures grant control rights to private oil companies. In other words, they enable mixed control over the oil upstream sector—both state and private control.

15 TNK-BP is Russia's third largest oil producing consortium.

16 Rosneft is 75.16 percent government-owned (PIW 2011).

17 AAR is a consortium of Russian oligarchs, reuniting three of Russia's leading investment, financial, and industrial groups: Alfa Group, Access Industries and Renova (AAR official website).

18 As will be explained in the operationalization section of the empirical analysis, an oil producing country needs to be neither oil exporting nor oil dependent. For more on this, please see Chapter IV.

19 It should be noted that Palacios (2001, 2002, and 2003) and Benton (2008) seek to explain energy policies in the Latin American region. Jones Luong and Weinthal (2001, 2006, and 2010) address energy development strategies in post-Soviet countries once they gained authority over their reserves—yet their interest lies more in rethinking the resource curse by examining the effects of these structures on institutions. See Chapter 3 for further discussion.

20 The two concepts of nationalization and expropriation are used interchangeably—as synonyms in the existing scholarship.

21 As Kobrin shows, "in practice, it may be difficult to distinguish coercion from the bargaining posture of an investor" (1984b, 330).

22 The Middle East produces more than any other region (BP 2011).

23 While there is an abundant literature in corporate finance on "firm control" (i.e. decision-making, power and influence within firms), there is no scholarship on "sectoral" or "industry control." Given the data scarcity, Jones Luong and Weinthal in their work from 2010 coded a categorical (four-value) variable which tries to distinguish between regimes which legally permit (or not) foreign oil companies to operate in the upstream in the presence/absence of ownership rights. In turn, both Benton (2008) and Palacios (2001, 2002 and 2003) use descriptive denominations of openness/closeness of the energy sector in their more qualitative, case study based analyses.

References

AAR Official Website—Alfa Group, Access Industries, and Renova, www.aar.ru/en/consortium/about-aar.html (status: January 3, 2013).

Al-Obaidan, Abdullah M., and Gerald W. Scully, "Efficiency Differences between Private and State-owned Enterprises in the International Petroleum Industry," in: *Applied Economics*, 24 (1992), 237–246.

Benton, Allyson, "Political Institutions, Hydrocarbons Resources, and Economic Policy Divergence in Latin America," in: *2008 APSA Annual Meeting Papers*, available at www.international.ucla.edu/economichistory/summerhill/benton1.pdf (status: February 21, 2013).

Black, Brian, *Crude Reality. Petroleum in World History* (Rowman & Littlefield Publishers, Lanham, MD, 2012).

Bortolotti, Bernardo, and Mara Faccio, "Government Control of Privatized Firms," in: *Review of Financial Studies*, 22 (2009), 2907–2939.

British Petroleum, *BP Statistical Review of World Energy*, June 2011, available at www.bp.com/assets/bp_internet/globalbp/globalbp_uk_english/reports_and_publications/statistical_energy_review_2011/STAGING/local_assets/pdf/statistical_review_of_world_energy_full_report_2011.pdf (status: February 27, 2013).

—, London, January 2012, available at www.bp.com/liveassets/bp_internet/global bp/STAGING/global_assets/downloads/O/2012_2030_energy_outlook_booklet. pdf (status: February 27, 2013).

Brunnschweiler, Christa, and Simone Valente, "International Partnerships, Foreign Control and Income Levels: Theory and Evidence," in: *CER-ETH Economics Working Paper Series*, 11/154 (2011), available at http://ideas.repec.org/p/eth/wpswif/11-154.html (status: February 22, 2013).

Ciarreta, Aitor, and Shahriyar Nasirov, "Development Trends in the Azerbaijan Oil and Gas Sector: Achievements and Challenges," in: *Journal of Energy Policy*, 40 (2012), 282–292.

Economides, Michael J., and Ronald E. Oligney, *The Color of Oil: The history, the money and the politics of the world's biggest business* (Round Oak Publishers, Wellington, UK 2000).

The Economist, *Argentina's Energy Industry: Fill 'er up*, 21 April 2012, available at www.economist.com/node/21553070 (status: February 27, 2013).

EIA (Energy Information Administration), *Country Analysis Brief: Mexico*, 2012(a), available at www.eia.gov/countries/analysisbriefs/Mexico/Mexico.pdf (status: February 27, 2013).

—, *Country Analysis Brief: Kazakhstan*, 2012(b), available at www.eia. gov/countries/analysisbriefs/Kazakhstan/kazakhstan.pdf (status: February 27, 2013).

—, *Country Analysis Brief: Uzbekistan*, 2012(c), available at www.eia.gov/EMEU/cabs/Uzbekistan/pdf.pdf (status: February 27, 2013).

—, *Country Analysis Brief: Turkmenistan*, 2012(d), available at www.eia.gov/EMEU/cabs/Turkmenistan/pdf.pdf (status: February 27, 2013).

—, *Country Analysis Brief: Russia*, 2012(e), available at www.eia.gov/countries/analysisbriefs/Russia/russia.pdf (status: February 27, 2013).

Eller, Stacy L., Peter R. Hartley, and Kenneth B. Medlock, "Empirical Evidence on the Operational Efficiency of National Oil Companies," in: *Empirical Economics*, 40 (2011), 623–643.

Elsas, Ralf, and David Florysiak, *Dynamic Capital Structure Adjustment and the Impact of Fractional Dependent Variables*, SSRN 1632362 (2010), available at http://papers.ssrn.com/sol3/papers.cfm?abstract_id=1632362 (status: February 22, 2013).

—, *Dynamic Capital Structure Adjustment and the Impact of Fractional Dependent Variables*, Revised version, SSRN 1632362 (2012).

ExxonMobil, *2012 The Outlook for Energy: A View to 2040*, available at www.exxonmobil.com/corporate/files/news_pub_eo2012.pdf (status: February 27, 2013).

Falola, Toyin, and Ann Genova, *The Politics of the Global Oil Industry: An Introduction* (Praeger Publishers, Santa Barbara, CA, 2005).

Financial Times, *Non-Public 150—The Full List*, December 14, 2006, available at www.ft.com/intl/cms/s/2/5de6ef96-8b95-11db-a61f-0000779e2340.html#axzz2M5hnMRYg (status: February 27, 2013).

—, *The New Seven Sisters: Oil and Gas Giants Dwarf Western Rivals*, March 11, 2007, authored by Hoyos, Carola, available at www.ft.com/intl/cms/s/2/471 ae1b8-d001-11db-94cb-000b5df10621.html#axzz2foc9GcFy (status: September 24, 2013).

—, *Rosneft to Pay $55bn in TNK-BP Takeover*, co-authored by Guy Chazan and Catherine Belton, October 22, 2012, available at www.ft.com/intl/cms/s/0/9776a77a-1c39-11e2-a14a-00144feabdc0.html#axzz2M5hnMRYg (status: February 27, 2013).

Jones Luong, Pauline, and Erika Weinthal, "Prelude to the Resource Curse Explaining Oil and Gas Development Strategies in the Soviet Successor States and Beyond," in: *Comparative Political Studies*, 34 (2001), 367–399.

—, "Rethinking the Resource Curse: Ownership Structure, Institutional Capacity, and Domestic Constraints," in: *Annual Review of Political Science*, 9 (2006), 241–263.

—, *Oil Is Not a Curse: Ownership Structure and Institutions in Soviet Successor States* (Cambridge University Press, New York, 2010).

Kobrin, Stephen J., "Foreign Enterprise and Forced Divestment in LDCs," in: *International Organization*, 34 (1980), 65–88.

—, "Political Assessment by International Firms: Models or Methodologies?", in: *Journal of Policy Modeling*, 3 (1981), 251–270.

—, "The Nationalisation of Oil Production, 1918–80," in: *Risk and the Political Economy of Resource Development* (Macmillan, London, 1984a).

—, "Expropriation as an Attempt to Control Foreign Firms in LDCs: Trends from 1960 to 1979," in: *International Studies Quarterly*, 28/3 (1984b), 329–348.

Krapels, Edward N., "The Commanding Heights: International Oil in a Changed World," in: *International Affairs*, 69/1 (1993), 71–88.

Mahdavi, Paasha, *State Ownership and the Resource Curse: A new dataset on nationalizations in the oil industry*, SSRN 1916590 (2011), available at http://papers.ssrn.com/sol3/papers.cfm?abstract_id=1916590 (status: February 22, 2013).

Marcel, Valerie, *Oil Titans: National oil companies in the Middle East* (Brookings Institution Press, Washington, D.C., 2006).

Marcel, Valerie, and John V. Mitchell, "How It All Started," in: Marcel, Valerie, *Oil Titans: National oil companies in the Middle East* (Brookings Institution Press, Washington, D.C., 2006), pp. 14–36.

Morse, Edward L., "Energy Breaks the Economic Rules," in: *SAISPHERE 2005*, available at http://legacy2.sais-jhu.edu/pressroom/publications/saisphere/2005/morse.htm (status: February 27, 2013).

OPEC—Organization of Petroleum-Exporting Countries, *Official Website*, www.opec.org (status: February 28, 2013).

Palacios, Luisa, *Explaining Policy Choice in the Oil Industry: A look at Rentier institutions in Mexico and Venezuela (1988–1999)*, PhD Thesis, Johns Hopkins University, 2001.

—, "The Petroleum Sector in Latin America: Reforming the Crown Jewels," in: *Science Po CERI Studies*, 88 (2002), available at www.sciencespo.fr/ceri/en/content/petroleum-sector-latin-america-reforming-crown-jewels (status: February 22, 2013).

—, "An Update on the Reform Process in the Oil and Gas Sector in Latin America," in: *Japan Bank for International Cooperation (JBIC) Working Papers*, September 2003.

Parra, Francisco, *Oil Politics. A modern history of petroleum* (I.B. Tauris, 2004).

PIW—Petroleum Intelligence Weekly, *PIW Top 50* (Energy Intelligence, December 2011).

Rice University, James A. Baker III Institute for Public Policy—*Project: The role of national oil companies in international energy markets*, available at www.bakerinstitute.org/events/the-changing-role-of-national-oil-companies-in-international-energy-markets (status: February 27, 2013).

Ross, Michael, *The Oil Curse: How petroleum wealth shapes the development of nations* (Princeton University Press, Princeton, NJ, 2012).

Thurber, Mark C., *NOCs and the Global Oil Market: Should we worry?*, Presentation Energy Seminar, Stanford University, February 6, 2012, available at http://energyseminar.stanford.edu/sites/all/files/eventpdf/Thurber%20energy%20seminar%20NOCs%2006Feb2012%20final_0.pdf (status: March 25, 2013).

Tordo, Silvana, Brandon S. Tracy, and Noora Arfaa, *National Oil Companies and Value Creation* (World Bank Publications, Washington, D.C., 2011).

United Nations General Assembly, *Resolution 1803 (XVII) of 14 December 1962 "Permanent Sovereignty over Natural Resources,"* available at http://untreaty.un.org/cod/avl/ha/ga_1803/ga_1803.html (status: February 27, 2013).

Victor, David G., David R. Hults, and Mark Thurber, *Oil and Governance: State-owned enterprises and the world energy supply* (Cambridge University Press, New York, 2012).

Victor, Nadejda, "On Measuring the Performance of National Oil Companies (NOCs)," in: *PESD Stanford University Working Papers*, 64 (2007), available at http://pesd.stanford.edu/publications/nocperformance (status: February 22, 2013).

Weinthal, Erika, and Pauline Jones Luong, "Combating the Resource Curse: An Alternative Solution to Managing Mineral Wealth," in: *Perspectives on Politics*, 4/1 (2006), 35–53.

Wolf, Christian, "Does Ownership Matter? The Performance and Efficiency of State Oil vs. Private Oil (1987–2006)," in: *Journal of Energy Policy*, 37 (2009), 2642–2652.

World Bank, *A Citizen's Guide to National Oil Companies* (World Bank Publications, The World Bank Group and The Center for Energy Economics at The University of Texas at Austin, 2008).

Yergin, Daniel, *The Prize: The epic quest for oil, money and power* (Free Press, New York, 1991).

—, *The Prize: The epic quest for oil, money and power*, with a new epilogue (Free Press, New York, 2009).

Yergin, Daniel, and Stanislaw Joseph, *The Commanding Heights. The battle for the world economy* (Simon & Schuster, New York, 1998).

2 Upstream sector policy in the oil industry

Policies emerge in an institutional framework. Understanding how institutions embed and change policies is of high relevance both to economists and political scientists. In reference to Dunning and Pop-Eleches, institutions are "sets of formal, rule-based constraints on the behaviour of individual and collective actors" (2004, 6). In this set-up, institutions are different from "sanctions, taboos, customs, traditions, and codes of conduct" (North 1991, 97), which create informal constraints on behavior. Despite this focused approach to institutions as formal constraints, this book acknowledges that informal constraints are relevant in any polity as they mould the basis for formal institutional arrangements.[1]

Along the same lines, Roland distinguishes between slow- and fast-moving institutions. Slow-moving institutions include social norms and beliefs (i.e. informal constraints) while fast-moving institutions are epitomized by political institutions like regime type, electoral rule, rules related to legislative bargaining or the degree of federalism (i.e. formal constraints). Compared to social norms, beliefs and values, which change continuously yet very slowly, political institutions are likely to change not at all or very little for extended time periods and then dramatically.

> An appropriate analogy is an earthquake: pressures along fault lines build up continuously but slowly, then suddenly provoke an earthquake that abruptly changes the topography of a given area. Slow-moving institutions are the equivalent of these tectonic pressures; fast-moving institutions are the equivalent of the topography.
>
> (Roland 2004, 117)

Within the framework defined by formal institutional constraints or "fast-moving institutions," economic policies are official actions taken by individuals and/or organizations (Dunning and Pop-Eleches 2004, 6)—ideally in such a way as to maximize the collective ophelimity[2] function (Tinbergen 1952, 1). In the narrow but most widespread sense of the term, economic policy[3] refers to the government. As will be discussed below, upstream sector policy comes under this category of government economic policy.

This chapter seeks to conceptualize upstream sector policy. To this end, the first part maps out the broader context in which upstream sector policies are chosen by discussing the institutional framework for policy-making as well as the range of commercial and/or non-commercial objectives pursued by oil producing countries. The second part theorizes the concept of upstream sector policy by delving into its dimensions and variants, and not least, addresses the tools through which upstream sector policy is "de facto" implemented in oil-rich countries. Given the scarce scholarship on upstream sector policy, such an endeavor is imperative before any further research is undertaken.

2.1 The political economy of the oil sector

In order to understand the design of the policies developed and pursued in the upstream sector, it is mandatory to gauge the context in which they emerge. For this purpose, this section looks into the political economy of the petroleum sector structures by addressing two main questions: first, *who sets the objectives for the petroleum sector?* and second, *what are these objectives?* They will be discussed in separate sections below.

2.1.1 The institutional set-up for policy-making in the oil sector

The overall objectives of the petroleum sector are defined at the highest levels of government. Notably, sectoral objectives need to be distinguished from operational targets. The former are factored into the petroleum policy, which has to be aligned with the national development policies. In theory, the crux of the petroleum policy is the maximization of government revenues under the constraints set by depletion rate, production quotas (if existent, e.g. under Organization of Petroleum-Exporting Countries (OPEC) quotas), and national development and/or social welfare policies. In turn, operational targets look into the technical implementation of the objectives set by the government in the petroleum sector and have to deal mostly with individual project-based decisions at different stages in the supply value chain (such as exploration, development, production, refinery etc.). Depending on the petroleum sector structure in place, such operational targets are left (to a larger or smaller extent) at the discretion of operators—most often, the National Oil Company (NOC) (Marcel 2006).

Generally, the ministry of petroleum is the government authority which sets the rules of the game in the petroleum sector. The level of technical and commercial knowledge accumulated by the ministry over time shapes the parameters of the relationship between government and the NOC and determines the type of policy-making process, i.e. government-dominated (top-down) or company-dominated (bottom-up) (Lahn *et al.* 2007, 20f.). More explicitly, when the ministry of petroleum has developed strong institutional capacity and also disposes of good knowledge of the sector's resources and investment needs, a top-down process is likely to prevail. There are situations whereby not the government but instead the state leader sets the objectives in the petroleum sector. In turn, in case of bottom-up processes, large information asymmetries between government and the NOC tilt the balance in favor of the latter whose power emerges primarily from its knowledge of the industry and inside business.[4]

Depending on the political system, decisions about the petroleum sector might be subject to public debate and/or parliamentary approval in one form or another. More specifically, in parliamentary and congressional systems, the ministry of petroleum holds objective-setting responsibility and is accountable to a legislature. Such are the cases of Brazil, Canada, Iran, Kuwait, Malaysia, Norway, the United States, and the United Kingdom. However, the span of necessary parliamentary approval might range from general legislation and government policy in some countries to discrete decisions on agreements with IOCs and the annual NOC budget in others. Furthermore, there are countries such as Mexico, Venezuela or Colombia, where the oil workers' unions are very vocal in sector policy-making (Audinet *et al.* 2007). In this regard, the Chatham House Report (Lahn *et al.* 2007, 24) cites the example of the Mexican Oil Union,[5] which played an instrumental role in the nationalization of the petroleum sector and the creation of Pemex.

Despite increasing the legitimacy of the adopted decision, parliamentary and public debate has not always led to optimal results from an efficiency point of view. There are cases where the involvement of multiple institutions in the decision-making process over matters requiring complex information and technical competence has been conducive to indefinite delays, has distorted the clarity of the industry's objectives and not least, has arguably forced poor environmental and commercial policies for the petroleum sector. The National Assembly of Kuwait in debating Project Kuwait (Abdulla 2010; Herb 2009), the Iranian Majlis in developing the terms of the buy-back contracts (Audinet *et al.* 2007, 57ff.), and the Russian Duma in deciding over potential production-sharing agreements on a project-by-project basis (Janosz *et al.* 1995) are just a few examples.

By contrast, in centralized political systems the decision-making process is hierarchical and dominated by the minister of petroleum sitting on the NOC board and/or the state leader at the apex of a higher governing council with regulatory powers, known as "Supreme Petroleum Council" (Lahn *et al.* 2007, 25). There is more room for change as a consensus is more readily reached in this type of system where the head of state and/or the government are to outline the roles and objectives, thereby shunning public or inter-agency debate about the policies taken in the petroleum sector (Abdulla 2010; Herb 2009).

Apart from the need for accountability (or lack thereof) specific to different political systems, the type of NOC also defines the decision-making remit of the government in the oil sector. The most autonomous

NOCs, which retain responsibility over operational decisions, follow the "corporatized model," whereby in accounting and financial terms the NOC is on a par with any IOC. More specifically, the NOC collects the revenues from oil operations and subjects its remaining profit to royalty, taxes and dividends. Yet the investment plans require approval from the government, which is the main (at times, the exclusive) company shareholder. This corporatized model is found in Saudi Aramco (Saudi Arabia), Sonatrach (Algeria), ADNOC (the United Arab Emirates (UAE)), Petronas (Malaysia), Petrobras (Brazil), and Statoil (Norway). At the opposite pole, there is the "pure budget model" where the government takes over both policy-making and operational decision-making in the petroleum sector. A given budget is allotted each year to the NOC which is supposed to fulfill strictly imposed project-based targets under the financial constraints. Yemen Petroleum Company is a typical case of this model. Finally, the "hybrid model" of NOC identified in Kuwait Petroleum Corporation and National Iranian Oil Company combines budget model conditions for upstream activities with corporatized model status for other activities[6] (Audinet *et al.* 2007).

Strong demarcation of functions and responsibilities is of major importance for the clarity of goals and objectives (see section 2.1.2 below). In fact, the Chatham House "Report on Good Governance of the National Petroleum Sector" (Lahn *et al.* 2007) distinguishes four key governance functions: "policy-making," "strategy-making," "operational decision-making," and "monitoring and regulation." This categorization of responsibilities ties into the scholarly debate about the administrative design of the petroleum sector and its impact on performance. The separation-of-functions approach, epitomized by the "Norwegian Model," was long professed as "a 'best practice' of sorts" (Thurber *et al.* 2010, 2011). In this case, the government functions are separated into: "policy," "regulatory," and "commercial." The "policy" function corresponding to the "policy-making" function in Chatham House's jargon appeals to the ministry of petroleum, which sets the objectives for the oil sector while bearing in mind broader macro-economic considerations. The "policy-making" function is not clearly distinguishable from the "strategy-making" function, which looks into how the petroleum sector will deliver national policy objectives. Yet it should be different from "operational decision-making," which deals with managing more short-term operations on the ground within the strategic framework. This is also called "commercial" function and in the "Norwegian Model," it is assigned to the NOC. Finally, "monitoring and regulation" or in other words, the "regulatory" function is vested into an autonomous oversight agency like the Norwegian Petroleum Directorate or more generally, the Supreme Petroleum Council, which should ensure that policies are being implemented and national objectives are pursued and attained.

The important separation is that of the commercial function from the policy and regulatory ones or synonymously, of operational decision-making from policy-making, monitoring and regulation (Thurber *et al.* 2010, 2011).

While the "Norwegian Model" may be no panacea for sectoral performance, petroleum-rich countries with a strong policy-making institution, i.e. ministry, and a separate operator, i.e. NOC, with decision-making power on the ground have proven to be better off than countries where functions have been muddled and chores have overlapped (Boscheck 2007).

All in all, this section tried to demarcate the political economy of the petroleum sector as well as the spate of functions and responsibilities of different institutions. The attention will now shift to the actual objectives pursued in the oil industry.

2.1.2 The commercial versus non-commercial ends of the range

Ideally, in reference not just to the United Nations General Assembly (UNGA) Resolution 1803 but also to the Natural Resource Charter (2010) assembled by an independent group of experts led by Paul Collier, "[t]he development of a country's natural resources should be designed to secure the *greatest social and economic benefit* for its people" (cf. Precept 1). As shown above, transposing this high-end goal into a feasible sectoral policy with realistic objectives comes under the prerogatives of the ministry of petroleum/government.

Yet the challenge of the government in countries which have created a state-owned enterprise (SOE) for oil and gas operations results from its dual role—that is, "the role of *sovereign owner* (on behalf of the people) when setting energy policy and the role of *company shareholder* (profit and value-creation) when engaged in commercial decision-making" (Lahn *et al.* 2007, 10). This duplication of roles spurs in turn "a conflict of interests between the allocation of expenditures required by the petroleum sector to deliver on geology, and the ones required by the government to deliver on non-oil development" for the benefit of the people (Audinet *et al.* 2007, 16). In this sense, the range of objectives for the petroleum sector spans from commercial to non-commercial or more generally, from economic to non-economic ends.

Arguably, the main sectoral objective for each petroleum-rich country is of *commercial* nature and can be defined as the maximization of revenues accruing to the state given the constraints set by depletion rate and production quotas (as already introduced above). Ideally, depletion policy should be a function of "allocative efficiency," which in abstract terms "requires that production levels match those desired by society" (ibid., 18). More explicitly, allocative efficiency includes a technical and a developmental side to it. The former considers the optimal depletion rate to be one which maximizes the recovery factor, which is determined by the production speed from the reservoir.[7] As part of "good oilfield practice" at the micro-level, these decisions are project-based and need to take into account the geological characteristics of the oilfield and reservoir/s. In turn, the latter regards the optimal depletion rate as the one which maximizes the country's wealth both now and in the future. This taps into the economic rationale applied by the government in developing the country's oil reserves. That is, a high development rate involves high revenue inflows in the present and near future, which is a valid policy choice as long as the revenues are productively used and invested with a long-term vision. A low development rate means keeping resources underground, postponing their extraction for times when conditions might be more favorable (technology-wise or price-wise) and thereby, reducing the potential to waste oil revenues in the short run. As recommended by the World Bank (Audinet *et al.* 2007, 19), at the macro-level the key criterion for the development rate of a country's oil reserves should be "the capacity of the country to productively invest the revenues either in developing the economy or in accumulating assets for future generations free from fear of corruption and theft."[8]

In framing and deciding over optimal ways of achieving this commercial objective, the government needs to find a balance between the total size of oil export revenues from which the government's share is taken, and the size of the government share also known as the "government take." In this context, the government faces a trade-off between private sector involvement and government take. More specifically, private sector involvement may increase the size of total revenues but reduce that of the government take, whereas barring private companies from the petroleum sector, especially the upstream, might increase the government take but shrink revenue growth if insufficient investments are made.[9]

Moreover, the petroleum sector is expected to deliver not only "on geology" by maximizing revenues while minimizing costs and optimizing production given depletion rate, but also "on (economic) development" (ibid.). In other words, it is expected that the petroleum sector has a wider positive impact at the national level by contributing to economic growth, development of non-oil economy and social welfare. Over collection of petroleum export revenues, these goals could be then pursued by the government through different policies. However, oftentimes the government lacks institutional capacity in mineral-rich countries and thus prefers to outsource some of these responsibilities by setting *non-commercial* objectives for the petroleum sector. These non-commercial objectives inform and build the so-called "national mission of the NOC" (ibid., 14) and analogously, the corporate social responsibility (CSR) of IOCs.

There are several tools through which non-commercial goals can be pursued in the petroleum sector. These can be split into two broad categories: forward and backward linkages. *Forward linkages* involve the provision of oil products for the domestic market. On the one hand, this depends on the domestic refinery capacity and/or the infrastructure capacity to import oil products. On the other hand, the willingness of operators in the petroleum sector to embark on forward linkages is strongly determined by the domestic pricing policy. In most oil producing countries, petroleum and its derivatives are highly subsidized on domestic markets, which inevitably leads to energy efficiency issues and financial losses. In this sense, domestic energy reforms have become imperative in e.g. Saudi Arabia, Kuwait, Iran and other MENA countries (Abdulla 2010; Cordesman 2003; Dincer and Al-Rashed 2002; Lahn *et al.* 2007; Mehrara 2007; Audinet *et al.* 2007 and also, BBC August 31, 2012), where governments have made strenuous endeavors to lift subsidies (at least, partially) while preventing dissent among the population. Depending on who bears the costs of subsidies, whether the petroleum sector (i.e. the operating companies; in most cases, the NOC though) or the central government, resistance can be encountered from the side of operators to provide these forward linkages.

Backward linkages can be reunited under the umbrella of "social value creation," which "refers to the creation of benefits or reductions of costs for society in ways that go beyond the maximization of the financial return on investment derived from the exploitation of the resource" (Tordo *et al.* 2011, xi f.). Backward linkages in the specific context of the petroleum sector encompass three main dimensions: (1) use and development of local labor force or in other words, creation of employment opportunities for nationals; (2) development of local products and services; and (3) creation of spill-over effects on the rest of the economy through investments in training and education of local employees or through transfer of technology. Clearly, there were cases over time where the scope of backward linkages went beyond these dimensions and operators in the petroleum sector were expected to build roads, hospitals, schools (e.g. see Eweje 2007, 2010 on Nigeria; Lippman 2004 and Vitalis 2006 on Saudi Arabia). Backward linkages, known to observe local content policies, are likely to "create distortions, inefficiency, and, in some cases, even corruption" (Tordo *et al.* 2011, 9).

Stark non-commercial objectives can turn into a burden for foreign investors and keep them away. While beneficial for the economic development of the host country, forward and backward linkages generally run counter to the profit-maximizing objective of the IOCs. Consequently, despite a legally permissive ("de jure") framework for foreign investments in the upstream sector, the conditions created by the host country (in view of its objectives) may well chase the investors away. This eventually explains why there might be a wedge between "de jure" and "de facto" foreign presence in the oil upstream sector of producer countries.

2.2 Upstream sector policy

In this section, the focus will turn to the concept of upstream sector policy. What is it? How does it relate to control? What about ownership? How is this policy implemented in practice? What are its tools? All these questions are to be dealt with in the following. To the best of my knowledge, there is no other work which seeks to conceptualize and define this type of economy policy. The World Bank's study (Tordo *et al.* 2011) presents an overview of the petroleum sector value chains, yet industry participation, licensing and petroleum contracts in the upstream as well as taxation are only briefly addressed.

Though a relevant question to the field of international political economy (IPE)—i.e. who controls the oil exploration and production sector?—it has so far elicited hardly any answer.[10] While the Corporate Finance literature abounds in works on "firm control," the scholarship remains completely silent on "sectoral control" or "industry control," let alone in the oil upstream.

To fill this gap, this section first seeks to capture the substantive content of upstream sector policy. In other words, before examining what shapes an upstream sector policy or in more operational terms, what the determinants of diverse control structures in the oil upstream sector are, we first need to understand what upstream sector policy and oil upstream industry control (or synonymously, oil upstream sectoral control) actually mean. For this purpose, the ownership-control dyad is presented and possible combinations of ownership and control as in variants of upstream sector policy are introduced. Second, the economic implications of state control in the upstream sector are addressed. Third and finally, the instruments through which upstream sector policy is implemented in practice are presented.

2.2.1 Conceptual insights

Within the scope of this book, *upstream sector policy* refers to sector participation—more specifically, state versus private sector participation in the oil production sector.[11] Notably, "[t]he state's decision—regarding *sector governance* and *sector participation*—are fundamentally interconnected and jointly affect value creation" (Tordo *et al.* 2011, 39, *my italics*). While *sector governance* is the result of the institutional arrangements present in the petroleum sector (discussed in section 2.1.1), the upstream sector policy or synonymously, the state's decision over *sector participation* in the upstream can be seen as the consequence of the objectives set for the sector (see section 2.1.2).

Sector participation is related to two core elements: ownership and control. *Ownership* stands for "a set of relations among multiple claimants to the benefits derived from the exploitation of the asset in question" (Jones Luong and Weinthal 2010, 9f.). Regardless of its type—whether public, private, foreign or domestic—ownership builds a form of exclusion. More explicitly, it grants "the sole right to exploit—and thereby to have direct access to the proceeds from an asset to which society assigns worth" (ibid., 10; see also Schlager and Ostrom 1992). Whoever holds the ownership rights (or synonymously, property rights or sovereign rights) over the petroleum assets has in fact the eminent domain rights—namely, the right to tax (in cash or in kind), the right to revoke an already granted or conceded right, and the right to police in the sense of regulating (Mommer 2002, 98). Importantly, in the case of upstream sector policy, ownership refers to reserves and not to operations.

Within this research, *control* means "operational control"—in reference to Boscheck's (2007, 373) schematization of the governance of oil supply—and it refers to the production activities performed in the upstream sector, which may involve no, partial or complete contracting to private oil companies. Sectoral control rights in the upstream grant decision-making power over the amount of investments, production volumes, and prices (Mommer 2002, 97).

Since the 1970s, state ownership over petroleum reserves has prevailed among oil producing countries.[12] Based on Jones Luong's (2010) dataset[13] on variation in ownership structure for the time period 1900 to 2005, there are only some scattered cases of private foreign ownership (e.g. Argentina, Bolivia, Equatorial Guinea, Gabon, Kazakhstan etc.) and private domestic ownership (e.g. Russia). Also for the Middle East and North African (MENA) region,[14] the default status since nationalizations has been state ownership, with the exception of North Yemen,[15] Yemen,[16] Algeria,[17] and Sudan.[18] This shows the fact that across the world, variation in upstream sector participation or synonymously, in upstream sector policies, mostly lies with the second dimension: control (read: operational control). The same observation was made by Brunnschweiler and Valente (2011, 3).

Control can be twofold: (1) state control, when the upstream operations are undertaken by the NOC; and (2) private control, which is most of the time foreign control, meaning that foreign oil companies or IOCs run (some of) the upstream activities while deciding over investment flows, production quotas and pricing. Notably, private control seldom reaches 100 percent in the upstream—thus, private control actually needs to be conceived as mixed control (i.e. both private and state control).

The combination of different forms of ownership (of reserves) and control (of production) can lead to three main forms of sector participation, i.e. to three basic upstream sector policies, as illustrated in Table 2.1.

A fourth form, made of private ownership and state control, has been practically non-existent. The most recurrent forms of sector participation are the first two, whereby the state holds the ownership of reserves and either controls the production through the SOE (variant 1) or grants control rights to private companies to different extents (variant 2).

Table 2.1 Combinations of ownership and control

Variant	Ownership	Control
(1)	State	State
(2)	State	Private
(3)*	Private	Private

* Jones Luong and Weinthal (2010) split the third variant into: private domestic ownership and control, and private foreign ownership and control.

Along the same lines, Mommer (2002) discusses the differences between non-proprietorial and proprietorial governance of the upstream sector. The former, indicative of the third variant in the table above, aims at enabling a free flow of investments into exploration and extraction while the guiding criterion is profitability of investments. The centerpiece of non-proprietorial governance is a fiscal regime based on excess profit taxation. By contrast, proprietorial governance, representative of the first two variants in Table 2.1 above, is wary of purely commercial objectives. That means that proprietorial governance confers control only if expected benefits (in cash or kind) are satisfactory to both investors and natural resource owner (i.e. state). The extreme variant of proprietorial governance is full nationalization of the oil upstream sector depicted by the first variant of upstream sector policy in the table.

While in the case of complete private control, decision-making is considered as fully pertaining to private companies, it is acknowledged that the state might exercise indirect control over the decisions of foreign companies and even more so, of private domestic companies. Throughout the world this can be attained through different leveraging mechanisms (e.g. pyramids, dual-class shares, cross-holdings)[19] and forms of political influence (e.g. nomenklatura system, investment approval system, distribution monopoly),[20] which (if existent) will be explored in the empirical analysis within the case studies.

2.2.2 State control versus economic performance

Drawing on the previous section, the extreme form of closure of the upstream sector is "state control," which automatically involves "state ownership" as well (henceforth: state control). Economically, there is wide agreement among scholars and practitioners alike that state control or in other words, an upstream run through a SOE, is less efficient than private control. This is frequently blamed on poor technical capabilities, and inadequate human resources policies in NOCs (Al-Mazeedi 1992; Gochenour 1992). The World Bank (Tordo *et al.* 2011) claims that NOCs fell behind in technological competency shortly after nationalizations, when IOCs invested most of the windfall profits resulting from high oil prices into research and development of new technologies aimed at cutting costs and increasing productivity. As for human resources, NOCs were often overstaffed, paid very high wages as compared to the rest of the economy, and were accused of recruiting based on patrimonial considerations rather than meritocracy (Al-Mazeedi 1992; Waelde 1996). Even though these conditions have (relatively) changed in the meantime, private oil companies continue to outperform national oil companies in operational terms (Al-Obaidan and Scully 1992; Eller *et al.* 2011).

In this sense, Victor (2007) finds that private oil companies are nearly one third better at converting reserves into output and generate considerably more revenue per unit of output. Wolf (2009, 2650) also points out that NOCs and in particular, OPEC NOCs are found to produce a much lower annual percentage of the upstream reserves than private companies. In contrast to other scholars, he argues that this might be caused by a more conservative depletion policy, a systematic overestimation of reserves, or by a mix of the two, and not necessarily by lower efficiency of state-owned firms. Contradictorily, he concludes that ownership effects clearly exist in the oil and gas industry and "a political preference for State Oil usually comes at an economic cost" (ibid., 2651).

In the literature on privatization, there are two main explanations for the poor (or relatively poorer) performance of SOEs, which come under the ownership-versus-environment debate. One explanation is that SOEs are more inefficient because of principal-agent problems. (Boycko *et al.* 1996; Ehrlich *et al.* 1994). In this case, the government of the oil producing country is the principal, and the NOC is the agent (Thurber *et al.* 2010, 2011; Stevens and Dietsche 2008). Because of information asymmetries due to the scale and technological complexity of the industry, the government loses ground in favor of the NOC which, in turn, is fairly free to pursue its own interests (Nore and Turner 1980; Van der Linde 2007; Vernon 1971). The paradox is that historically, NOCs were primarily created for the oil producing countries to regain sovereignty and control over national resources to the detriment of the IOCs (Fadhil J. al-Chalabi quoted in Mahdavi 2011, 11); yet this has been (partly) hijacked by the NOCs as they have not relented full access to the information the government wished for. Instead, "NOCs often capitalize on the principal–agent relationship and information asymmetries between the domestic government and itself" (Tordo *et al.* 2011, 28), and become major players on their own (Stevens 2005).

More concretely, NOCs (as a form of SOEs) are controlled by bureaucrats, who have widely different goals from the social interest. Bureaucrats controlling state firms often cater to special interest groups (such as public employee trade unions), which can help them with elections, and are at best indirectly concerned about profits, which are beyond their control as they flow into the government budget (La Porta *et al.* 1998; Shapiro and Willig 1990). It is noteworthy that, "they have virtually complete power over these firms, and can direct them to pursue any political objective. State ownership is then an example of concentrated control with no cash flow rights and socially harmful objectives. Viewed from this perspective, the inefficiency of state firms is not at all surprising" (Shleifer and Vishny 1997, 768).

The opposite camp of the ownership-versus-environment debate contends that "ownership per se does not determine firm performance" (Bartel and Harrison 2005, 136); rather, it is the different environments in which public and private enterprises operate that explain the efficiency gap.

In this regard, Pinto and van Wijnbergen (1995) as well as Bertero and Rondi (2000) point to the soft budget constraint regime as the government provides additional resources, or when needed, bails SOEs out. Another related point to the environment explanation is that compared to private-sector enterprises, SOEs are likely to face different degrees of internal and external competition (Bartel and Harrison 2005; Mommer 2002; Nickell 1996). Competition enables comparisons of managerial and operational performance, encourages innovation, and creates a disciplined milieu in which companies need to fight for market share and against bankruptcy (Pollitt 1995; Tordo *et al.* 2011).

Based on evidence from Brazil's oil industry, it turns out that even the threat of competition can improve the performance of state-owned companies. After the reform of 1995, when Brazil ended the legal monopoly of Petrobras and set out the liberalization process of the oil sector, cf. the Petroleum Law of 1995, the company faced the prospect of competing against lower cost producers and losing their monopoly over the Brazilian market. On these premises, it renounced pursuing non-economic goals as it was granted more freedom by the government to manage its affairs. The threat of competition, even in the absence of actual competition, led to a considerable increase in the productivity of Petrobras, with labour productivity growth tripling as compared to the period prior to 1995 (Bridgman *et al.* 2011).

The positive effect of competitive pressure on productivity growth is evidenced in other sectors as well, ranging from iron ore mining (Galdon-Sanchez and Schmitz 2002; Schmitz 2005; Schmitz and Teixeira 2008) to the airline industry (Goolsbee and Syverson 2008).[21] Reuniting the individual effects of competition and ownership structures, Grosfeld and Tressel's study claims that these are in fact complements and not substitutes. That is, competition and "good" governance, which should be conceived as a function of ownership concentration, reinforce each other. On the basis of the economic performance data for approximately 200 firms listed on the Warsaw Stock Exchange over the period 1991 to 1998, the scholars conclude that "[c]ompetition affects productivity growth in firms characterized by 'good' corporate governance. In firms with 'bad' corporate governance the effect of competition is insignificant" (2002, 546).

In short, it appears that state control has its economic downside. Though not a panacea, more competition or at least, threat thereof, can lead to an increase in economic performance of national oil companies and more generally, state-owned enterprises. Yet a question remains unsolved: if state control underperforms economically, why is it still preferred in oil producing countries? Answers to this question will be provided in the next section, which sheds light on the pluses and minuses of different upstream sector policy tools.

2.2.3 Upstream sector policy tools

There are several tools through which upstream sector policy is implemented. Since creation of an NOC is only one of these tools, it is important to ensure coherence and coordination between the NOC's role and other policy tools. In the following, the reasons for NOC creation will first be presented. This should clarify how, for oil producing countries, losing at the economic end may mean gaining at the political end. Second, the discussion of other policy tools—namely, legal arrangements and fiscal systems—is meant to elucidate why in producer countries state control is preferably exercised through an NOC rather than indirectly through legal and fiscal engagements with private oil firms.

2.2.3.1 NOC creation

There is large variation among NOCs with regard to the level of competition faced in the markets where they operate, their business profile along the petroleum sector value chain, and the degree of commercial orientation and internationalization. To map out this diversity across NOCs, PFC Energy (2011), an international energy consulting company based in Washington DC, proposed the following typology:

(1) *"Façade/mailbox"* (e.g. NNPC in Nigeria), with no or limited capabilities and no actual operations. The role is confined to creation of employment and representation of the government in the petroleum sector in relations with foreign oil companies.

(2) *"Statist bureaucracy"* (e.g. PDVSA in Venezuela, LNOC in Libya, Sonatrach in Algeria, Sonangol in Angola, GNPC in Ghana), which provides patronage for the ruling elites, yet it has no broad contributions towards economic development.

(3) *"Development bureaucracy"* (e.g. CNPC, Sinopec—both in China, Saudi Aramco in Saudi Arabia, QP in Qatar, Rosneft in Russia, KMG in Kazakhstan, KPC in Kuwait, ADNOC in Abu Dhabi, PEMEX in Mexico, ONGC in India, NIOC in Iran, Pertamina in Indonesia), which ensures revenues to the government, provides domestic subsidies for fuel, and facilitates broader socio-economic development.

(4) *"Public entrepreneur"* (e.g. CNOOC in China, Petronas in Malaysia, Petrobras in Brazil, Ecopetrol in Colombia, PTT in Thailand), which supports the industrialization and growth efforts in the respective countries as they are granted more autonomy in the pursuit of commercial goals both domestically and globally.

(5) *"Privatized/competitive"* (e.g. Statoil in Norway), which invests in efficiency and development of technology and technical expertise,

enjoys full operational capacity, and is rather free to embark on business opportunities overseas.

These different types of NOCs reflect the structures of the state and comply with its needs. They evolve and regress together; otherwise, the NOC may be dissolved. The organizational capacity of the NOC thus corresponds to the political capacity of the state. More specifically, "façade/mailbox" NOCs with neither organizational nor operational capacities are likely to be found in quasi-states with unstable regimes fighting for state survival (like Nigeria), "statist bureaucracies" in extraction states ruled by a coherent elite with monopoly over the petroleum revenues (like Venezuela, Algeria, Angola, Libya), "development bureaucracies" in allocation states which, by comparison with extraction states, pursue a more extensive range of state services and subsidies for the population (e.g. Saudi Arabia, Qatar, Abu Dhabi in the UAE, Kuwait, Indonesia), "public entrepreneurs" in industrializing states with globally competitive industrial sectors and a diversified tax base (such as China, Malaysia, Brazil, Thailand), and finally, "privatized or competitive" NOCs in globalized states with well-established political institutions and a market driven economy (such as the case of Norway). Not only the operational autonomy, technical capacities and degree of commercial orientation rise from a "statist bureaucracy" to a "privatized/competitive" NOC, but also the ability to shape a long-term strategic vision and become international.

To control for the great diversity among NOCs, this book focuses on the NOCs of oil producing countries, which play a significant (if not dominant) role in the domestic petroleum industry. This approach is similar to the one adopted by the World Bank's study (Tordo *et al.* 2011). Summarized under the economic cost of state control (see section 2.2.2), NOCs-related setbacks have become evident. On these grounds, there should be a number of reasons why oil producing countries have decided to set up a NOC. These are addressed in the following.

First, the historical context in which most NOCs were established needs to be considered. Several of them are the result of asset nationalizations of foreign oil companies which were perceived to be ruled by imperialistic goals, detrimental to national interests and beneficial to foreign governments (Grayson 1981; Hartshorn 1993; Stevens 2008; Tordo *et al.* 2011). Restoration of control over mineral resources was regarded as a national mission for oil producing countries, even more so given the serious frictions emergent in the historical relations between host governments and IOCs (e.g. in Iran, Iraq—see Mitchell 2009, 2011; for Saudi Arabia—see Al-Rasheed and Vitalis 2004, Vitalis 2006) and the adoption of the UNGA Resolution 1803 (XVII) of December 14, 1962, "Permanent Sovereignty over Natural Resources."[22] Some scholars further identify an element of mimicry across countries in setting up an NOC since "creating symbols of independence became quite fashionable in the post-colonial world"

(Tordo *et al.* 2011, 22). Also, the proliferation of NOCs after World War II through the late 1970s may be tied to the prevailing political ideology of the time when the state was seen as a benevolent and necessary player in the socio-economic arena.

A second reason for setting up an NOC is the high relevance of the petroleum sector to the domestic economy in these oil producing countries. Most of them, in particular in the developing world, are "rentier" economies (Beblawi and Luciani 1987; Luciani 1990). While there are several contested measures for the level of "rentierism," they are all meant to capture the degree of national dependency on the petroleum sector. The level of dependence may create stronger (or less so) incentives for state control in the petroleum industry.[23]

Third, state control following NOC creation can bring political gains for the country both internationally and domestically. Internationally, control over petroleum resources boosts the bargaining position of the oil producing country among other states and can secure financial, political and military advantages (Klare 2004). Domestically, it provides the government with more technical and commercial information about the petroleum sector and also, with large(r) decision power over depletion policy and resource development, investment flows, prices, and scope of activities[24] (Myers Jaffe and Elass 2007; Wolf and Pollitt 2009).

A fourth reason for NOC creation refers to the benefits of state control for wider socio-economic issues through more direct involvement. These benefits have been discussed at large in section 2.1.2 under non-commercial objectives and comprise both forward and backward linkages.

Last but not least, NOC establishment and thereby, state control over the petroleum sector, can boost national pride, which ties into the "resource nationalism" thesis (Bremmer and Johnston 2009; Mitchell 2009; Vivoda 2009; Woodhouse 2006). In Stevens' opinion, resource nationalism "can be driven by a concern that the IOCs are taking too large a share of the cake. By a perception that the resource will be needed for domestic uses or that the potential customers are somehow 'unworthy'" (2008, 6). Consequently, the NOC is supposed to restore the justice over the petroleum resources.

All in all, at the macro-economic level, NOC creation can be justified in manifold manner based on: the historical context of the NOCs' proliferation, the economic relevance of petroleum revenues to the national economy of many oil producing countries, international political gains and state leverage over operational decision-making in the national petroleum industry, broader benefits for the economy and society at large, and national pride sentiments.

For the government in an oil producing country, the NOC is meant to ensure transparency about the workings of the petroleum sector. Away from the principal-agent problem between the NOC and government, the former should provide the latter with the necessary information to make informed decisions about existing oil reserves, depletion rates, the need for foreign technical expertise and the allocation of proceeds. The NOC can thus be seen as a direct lever of the government in the petroleum sector. Even when information asymmetries might arise, they are likely to be less large than between private oil companies and host producer governments (Ross 2012, 240f.). In the absence of an NOC, other policy tools are hands-off exercises of control by the state, and seriously raise the question of effectiveness in ruling the oil sector and taking the fair share of oil revenues. These will be discussed in the following section.

2.2.3.2 *Other tools*

Legal arrangements in oil producing countries allow for all possible combinations of control: private control, state control, and mixed control (i.e. state and private). The fiscal components associated with these different legal arrangements generate the government take, which is indicative of the share of oil revenues pertaining to the producer government. The discussion of these different arrangements with their fiscal systems should elucidate why oil producing countries prefer to exercise control through an NOC in the oil upstream sector.

2.2.3.2.1 LEGAL ARRANGEMENTS

The legal framework for exploration, development and production of petroleum reserves in the oil producer countries is set in the constitution, hydrocarbon law and regulations. The constitution lays out the fundamental principles regarding private property rights and it may interdict private parties or foreigners to obtain property rights in general and mineral rights in particular. Furthermore, the constitution may confer authority to grant petroleum rights and regulate specific matters upon state or provincial governments (and agencies) rather than the national government. Framed at the parliamentary level, the hydrocarbon law puts forward the principles of law and clarifies the objectives set for the hydrocarbon sector. Those provisions which do not directly affect the principles of law and/or need periodic adjustments (such as technical requirements, administrative procedures, administrative fees etc.) are, in turn, pinned down in regulations (Tordo 2007, 7; 2009, 8).

Depending on the legal system, a government grants exploration, development and production rights by means of two primary legal arrangements: *concessionary* and *contractual*. Under a *concessionary* system, the private oil company is entitled to the crude oil produced (at the wellhead or the borehole), for which it pays royalties and taxes. Additionally, ownership of the exploration and production equipment permanently attached to the ground passes on to the state at the termination or expiry of the concession (whichever comes first).

By comparison, under the *contractual* system, the government retains title to the crude oil produced. Notably, there are two main kinds of contractual systems: *production-sharing contracts (PSCs)* and *service contracts (or service agreements)*. Under a PSC, the investor (read: private oil company) acquires ownership of its share of production only at the delivery point or export point, and not at the wellhead. More explicitly, the ownership of the crude oil remains with the host producer state (through the NOC) and follows the scheme of cost recovery and only afterwards, production split. In turn, under a service contract, the contractor never acquires title to the resource. The oil company is paid a service fee (typically in cash), which is fixed (i.e. flat fee), in the case of pure service contracts or linked to the profit in the case of risk service contracts. As for the ownership over the exploration and production equipment commissioned or placed-in-service, under contractual systems this is transferred to the state immediately. The only exception to the rule is leased or service company's own equipment (Bressand 2009; Johnston 2003; Tordo 2007, 2009).

Drawing on Johnston (2003, 28), the shift from concessionary to contractual systems has been mostly caused by psychological reasons.[25] Historically, mineral rights were granted by concessions, which transferred rights to petroleum development and production over a vast area and also, over a long period of time to the private oil company. The investor was conferred extensive control over procedures, timing and volumes, while the sovereign retained few rights except the right to receive a production-based payment (Tordo 2009, 9). In the concession era dating back to the second half of the nineteenth century, it was the major private oil companies known as the "Seven Sisters" which set the initial conditions for oil exploration and production throughout the world (Bressand 2009, 117f.). With falling oil prices and less revenues in the 1950s to 1960s, governments felt the urge to intervene in the oil market, which further culminated in massive nationalizations of the hydrocarbon sectors and eventually gave leeway to contractual systems.

A few countries still use the concession-type based regime (e.g. the United States, the United Kingdom, France, Ireland, Spain, Chad, Portugal, Australia, South Africa). However, the provisions of modern concessions have changed considerably with regard to the exploration area and duration coverage, relinquishment clauses and government participation at different stages in the work programme. Unchanged remains the fact that the private oil company bears all the risks and costs associated with the exploration, development and production of crude oil within the area under concession.

Similarly to a concession, a PSC grants an investor (an oil company or a consortium) the right to exploration, development and production of petroleum in a given area for a specified time period. The first PSC dates back to 1966 and was signed between IIAPCO and the Indonesian National Oil Company (now Pertamina). This initial contract encompassed the basic features, which have been preserved throughout time and continue to build a legal formula applied by many oil producing countries nowadays. These were: the title to the crude oil stayed with the state; Pertamina preserved management control while holding IIAPCO responsible for execution of operations based on contractual terms (including work programmes, budgeting); IIAPCO assumed all risks and costs, i.e. both financing and technological equipment required for the operations; the contract was based on production-sharing and not profit-sharing—consequently, after cost recovery of up to 40 percent of annual oil production, the rest of the production was split 65:35 percent in favor of Pertamina; finally, all the equipment acquired and imported by IIAPCO, besides service company equipment and leased equipment, came under the ownership of Pertamina (Johnston 2003, 29f.).[26]

Under a *service contract*, the contractor is paid a fixed or variable fee to carry out exploration and/or production activities in a given area within a given timeframe. Contrary to concessions and PSCs, "[t]he state maintains ownership of petroleum at all times, whether in situ or produced" (Tordo 2009, 10). Pure (non-risk) service contracts are very rare. This contractual arrangement implies that the contractor performs exploration, development and/or production services on behalf of the host government against a fee. The risks are borne by the state. This form of pure service contracts can be, for example, still found in Iran as buy-back contracts (Johnston 2003, 41f.; Tordo 2009, 11). In turn, in the framework set by a risk service contract, "the contractor bears all the risk but has the potential of profits" (Johnston 2003, 41), meaning that if the exploration, development and/or production activities are successful, the host government allows the oil company to recover the costs and also pays it a fee, which is a given percentage of the remaining revenues (thus, profit-based). This fee can be paid in cash or in crude, and it is furthermore taxed within the host country (Johnston 2003; Nakhle 2010).

In conclusion to the contractual arrangements, while under the initial concessionary system private oil companies had full control rights over the amount of investments, production volumes, prices and revenues, this has drastically changed with the contractual arrangements. PSCs allow for mixed control—that is, shared between the state (through the NOC) and private oil companies, yet in the case of service contracts state control is exclusive. The fiscal elements of each of these legal arrangements are further addressed.

2.2.3.2.2 FISCAL SYSTEMS

Petroleum fiscal regimes can be very complex and vary from country to country. This section will not go into any details; rather, it presents the key fiscal elements of concessions, production-sharing contracts and service contracts, through which the government take is raised in oil producing countries.

In reference to the "Oil and Gas Royalties" Report by the U.S. Government Accountability Office (GAO) from September 2008, the government take is the sum of all revenues in royalties, rents, bonuses, corporate income taxes and other fees, collected by the government from oil and gas operations (see also U.S. GAO 2007). "The terms and conditions under which the government collects these revenues are referred to as the 'oil and gas fiscal system'" (U.S. GAO 2008, I). The purpose of the oil and gas fiscal system of the United States, which could well apply to any oil producing country, is to "strike a proper balance between maintaining competitive investment conditions and providing an appropriate share of revenues to the public" (ibid., 8).

In other words, from the host government's point of view, the mix of fiscal elements should, on the one hand, maximize wealth from the national resources, and on the other hand, attract investments for appropriate levels of exploration, development and production activity. As Johnston puts it, "[t]he issue of the division of profits lies at the heart of contract/license negotiations" (2003, 5), which are in turn accommodated by different allocation systems.[27] Depending on the geological conditions and not least, political risks in the oil producing country, governments may demand a higher or a less high take from the oil operations.

Currently, the balance in most producer countries tilts towards flexible fiscal systems where the government take is a function of profitability. The most common method applied to establish a flexible fiscal system is based on sliding scale terms, which imply progressively smaller share of profit oil for the private oil company as production grows. This should ideally encourage development of both large and smaller fields, of both less and more mature fields and also, "create a framework that can honour the mutuality of interest between the host government and the contractor" (Johnston 2003, 17). Some contracts may link more than one variable to a sliding scale[28]—for example, cost recovery limits, profit oil, and royalties.[29] As opposed to flexible progressive systems, host countries can operate under regressive fiscal systems where the government extracts front-end loaded rents such as royalties (before profit, yet post-discovery) or at times, even pre-discovery such as signature bonuses. A neutral fiscal system is back-end loaded, which means that the rents raised by the host government are revenues-based (Johnston 2003). Nevertheless, completely neutral fiscal systems may exist in theory but hardly so in practice. Royalties are the common pendant of both concessionary and contractual systems (with the exception of service contracts).

The main characteristics of the three categories of legal arrangements will be now presented. In its most fundamental form, a *concessionary* system or a royalty-tax (R/T) system has three basic components: royalty, deductions (operating costs, depreciation, depletion and amortization, and intangible drilling costs), and taxes (Tordo 2009, 13). Historically, royalties have been the most important instrument of taxation in mineral-endowed countries.

They are attractive to governments since "they ensure an up-front revenue stream as soon as production starts" (Baunsgaard 2001, 10). The royalty rate, calculated as a percentage of production volume ("unit" or "specific" royalty) but more typically, of production value ("ad valorem" royalty), can deter investments if set too high because of increasing marginal costs for the contractor. To make a project look more profitable to oil companies, royalties may be determined on a sliding scale, the terms of which are negotiable, statutory or biddable. Furthermore, as compared to a contractual arrangement, in a concessionary system there are no cost recovery limits. As for the taxable income, this may be taxed at the country's basic corporate tax rate. Special incentive-based programs as well as special resource taxes may apply to encourage investments or alternatively, collect the fruits of profitable oilfields[30] (Tordo 2007, 9, 37f.).

A *production-sharing contract* has in turn four main components: royalty, cost recovery, profit oil and taxes (Johnston 2003). Similarly to a concessionary system, the royalty comes right off the top, is usually not cost recoverable but tax deductible. Notably, there are PSCs which do not encompass royalties. Yet the technical distinction from a concession is the cost recovery, which imposes limits on the amount of deductions to be taken in any accounting period. More explicitly, if operating costs and depreciation exceed the limit placed on the volume of production or revenues considered for cost recovery in a given accounting period, the balance is then carried forward and recovered later. The rest of revenues after royalty and cost recovery is called profit oil and is analogous to the taxable income in the concessionary system. Johnston further points out that, "[t]he terminology is precise because of the ownership issue. The term taxable income implies ownership that does not exist yet under a PSC. The contractor has nothing to tax—not yet" (2003, 32). Finally, taxes may apply or be taken over by the host government or NOC on behalf of the contractor in accordance with the contractual terms (Tordo 2009, 13).

As for the *service contracts*, the contractor receives a service fee from the host government upon execution of given exploration, development and/or production operations. This service fee is the correspondent of taxable income from the concessionary system or the profit oil from the PSC and is thus taxable. More specifically, corporate income taxes may apply, yet fiscal costs and rules for amortization and depreciation can be established in the contract. No royalty and also, usually no recovery cost limit apply (Johnston 2003; Tordo 2009).

To provide a comparative basis and a frame of reference, Johnston looked into 55 concessions (or royalty-tax systems) and 68 PSCs around the world, as of 1996. The statistics show that on average, in concessionary systems, 98 percent of costs were recovered by contractors in a given accounting period (de jure: no recovery limit) as compared to only 63 percent in PSCs. As for the government take, this was as high as 58 percent in concessions and 66 percent in PSCs (Johnston 2003, 72). Service contracts are not directly factored into this analysis but Johnston generally draws them closer to PSCs. Different legal arrangements with their corresponding fiscal elements thus bring about different benefits for contractors (private oil companies) and host countries.

Based on Johnston's (2003) statistics, it turns out that PSCs raise higher government take than concessions and are thus financially more advantageous for oil producing countries. This is further strengthened by the recovery limits for contractors—the lower the recovery limit, the higher the potential profit for the oil producing country. As long as the investment conditions remain attractive for private oil companies and they are willing to engage in this legal arrangement with its fiscal system (PSC or respectively, service contract), this can only show that the hands-on control exercised by the state through an NOC in the domestic petroleum sector enables the former to better negotiate with private oil companies. State control through an NOC grounds the host producer government more firmly in the realities of the oil sector and gives it more footing when employing the services of private oil firms. Eventually, this should explain why host producer governments prefer the NOC as a tool in their pursuit of control in the oil upstream sector.

All in all, this chapter sought to conceptualize the focus of this research: upstream sector policy. Following its contextualization, the concept was discussed in relation to its two sub-components (ownership and control), and finally featured in reference to the tools through which its hands-on or hands-off implementation is made possible.

Notes

1 Notably, "[s]uch a distinction may allow us to ask more clearly a range of questions about the interaction between formal rules (i.e. institutions) and social practices, expectations, norms, and beliefs: How do particular sets of beliefs or expectations undergird formal rules and give them their power to govern political, social, and economic life?" (Dunning and Pop-Eleches 2004, 6); in the case of this book, to ask how given sets of norms, beliefs and social practices embed the formal rules and give them power to shape economic policies like the upstream sector policy. These informal constraints will be particularly discussed in the case studies.

2 The concept was introduced by Vilfredo Pareto as a measure of purely economic satisfaction as opposed to the broader term of utility, which includes other dimensions of satisfaction (e.g. ethical, moral, religious, political).

3 An insightful definition of economic policy is provided by József Veress: "economic policy means the views, resolutions, regular decisions, acts of the state, which it applies for influencing the economy to achieve its social – political goals" (translated and quoted in Saghi 2002, 191f.). For a theoretical background to the concept of economic policy, please see Buchanan (1987), Dixit (1998), and Persson and Tabellini (2002).

4 This situation is one example of the principal-agent problem. For principal-agent theory in the NOC context, see Stevens (2008) and for a more general discussion of the principal-agent arrangements, see Jensen (1983).

5 Sindicato de Trabajadores del Petroleo de la Republica Mexicana (STPRM).

6 Please note that the efficiency of these NOC models is beyond the scope of this chapter.

7 This is also identified as "technical efficiency" and should be the result of engineering considerations.

8 For economic calculations regarding optimal depletion rate given oil price and discounted rate (plus inflation), see the Hotelling model of resource depletion (Khanna 2001; Tietenberg and Lewis 2000).

9 There are a number of policy tools to capture the rent (even) when private companies are involved in the petroleum sector (including the extraction segment). These are fiscal instruments connected with legal arrangements, and will be addressed in the next section with particular reference to the upstream. Notably, the distribution of oil revenues across government and society is beyond the scope of this chapter.

10 Jones Luong and Weinthal (2010) capture "de jure" sectoral control and ownership in the energy sector, which looks into the permissibility of the legal framework in a given producer country. This is in turn shaped by the political economy of the petroleum sector structures, as presented in section 2.1 above. Yet, "de jure" control is not necessarily indicative of the reality. Although the legal framework might allow foreign involvement in upstream operations, its unattractive provisions might not bring investors into the sector. Consequently, "de facto" control or operational control can draw a considerably different picture of the sector. It should be emphasized that this book discusses the legal provisions in place which set the rules of the game—however, it places its focus on "de facto" sectoral control.

11 Even though the upstream sector also includes exploration activities, it is impossible to gather systematic and comprehensive information on these. Most of the time, however, successful exploration is followed by extraction or production, which thus legitimizes the proposed definition.

12 In order for IOCs to have private ownership, the state has to make an explicit decision to sell the ownership rights to subsurface resources. The exception is the United States where the ownership rights belong to the titleholder of the place where (onshore) oil or gas is located.

13 I wish to thank Pauline Jones Luong for sharing her dataset with me.

14 This includes the following countries: Bahrain, Egypt, Iran, Iraq, Kuwait, Libya, North Yemen/Yemen Arab Republic, South Yemen/People's Democratic Republic of Yemen, Yemen (as of 1990), Algeria, Oman, Qatar, Saudi Arabia, Sudan, Syria, Tunisia, and the UAE.

15 North Yemen/Yemen Arab Republic—private foreign ownership: 1974–1989.

16 Yemen—private foreign ownership: 1990–2005.

17 Algeria—private foreign ownership: 2005.

18 Sudan—private foreign ownership: 1975–2005.

19 For more on this, please see Andrews-Speed *et al.* 2000, Bortolotti and Faccio 2009, Claessens *et al.* 2000.

20 For more on this, please see Berkowitz and Semikolenova 2006, Domjan and Stone 2010, Hanson 2009, Heinrich 2008, Locatelli 2006, Shleifer and Vishny 1997.

21 An exception to this is Zhang *et al.* (2001)'s study, which produces inconclusive results as to the impact of competition on the economic performance. Based on approximately 2,000 Chinese firms in 26 industries in the time period 1996–1998, their empirical analysis shows a significant positive correlation between international market competition and company efficiency; however, domestic market competition is negatively correlated with company efficiency. The

scholars explain the differential effect between international and domestic market competition by invoking the anomalies in the "Chinese socialist market economy" (Zhang *et al.* 2001, 12f.).

22 For this, please refer to Chapter 1.

23 Please note that petroleum rent maximization and more technically, the government take will be discussed below with reference to the fiscal systems.

24 Yet please be mindful of the principal-agent problem presented in section 2.2.2 above.

25 As a side remark, "[t]he term *concession* has a lot of negative connotations these days. *R/T system* is becoming the preferred terminology for those who care to be politically correct" (Johnston 2003, 18).

26 The oil industry has experienced the emergence of another contractual arrangement which is very similar to a production-sharing agreement. This is the so-called joint operating agreement (JOA). The most well-known form of this is the joint venture (JV), which differs in two aspects from a PSC: first, the risks and liabilities are held jointly by all the JV partners and not just by the private oil operator; second, financing of operations and share-out of revenues involve all the partners based on their percentage interest in the JV, as stipulated in the initial contract (Black and Dundas 1992).

27 Please note that allocation systems (e.g. open-door systems, licensing rounds with administrative procedures and auctions) are beyond the scope of this research. For this, please refer to Tordo 2009, 13–22.

28 The alternative to a sliding scale system, though bearing the same rationale, is a rate-of-return (ROR) system. On this, please see Johnston (2003, 42–51) and Baunsgaard (2001, 8f.).

29 All of these will be defined in the following in relation to legal arrangements.

30 In this sense, please refer to ringfencing in Baunsgaard (2001, 7) and Tordo (2007, 38).

References

Abdulla, Abdulkhaleq, "Contemporary Socio-political Issues of the Arab Gulf Moment," in: *LSE Kuwait Programme on Development, Governance and Globalisation in the Gulf States*, 2010.

Al-Mazeedi, Wael, "Privatizing National Oil Companies in the Gulf," in: *Journal of Energy Policy*, 20 (1992), 983–994.

Al-Obaidan, Abdullah M., and Gerald W. Scully, "Efficiency Differences between Private and State-owned Enterprises in the International Petroleum Industry," in: *Applied Economics*, 24 (1992), 237–246.

Al-Rasheed, Madawi, and Robert Vitalis, *Counter-Narratives: History, Contemporary Society, and Politics in Saudi Arabia and Yemen* (Palgrave Macmillan, New York, 2004).

Andrews-Speed, Philip, Stephen Dow, and Zhiguo Gao, "The Ongoing Reforms to China's Government and State Sector: The Case of the Energy Industry," in: *Journal of Contemporary China*, 9 (2000), 5–20.

Audinet, Pierre, Paul Stevens, and Shane Streifel, *Investing in Oil in the Middle East and North Africa. Institutions, Incentives and the National Oil Companies* (World Bank, Sustainable Development Department, Middle East and North Africa Region, 2007).

Bartel, Ann P., and Ann E. Harrison, "Ownership Versus Environment: Disentangling the Sources of Public-sector Inefficiency," in: *Review of Economics and Statistics*, 87 (2005), 135–147.

Baunsgaard, Thomas, *A Primer on Mineral Taxation* (International Monetary Fund, Washington DC, 2001).

BBC, *Arab Uprising: Country by country*, August 31, 2012, available at www.bbc.co.uk/news/world-12482678 (status: February 27, 2013).

Beblawi, Hazem, and Giacomo Luciani, *The Rentier State* (Routledge Kegan & Paul, London, 1987).

Berkowitz, Daniel, and Yadviga Semikolenova, "Privatization with Government Control: Evidence from the Russian Oil Sector," in: *William Davidson Institute Working Papers*, 826 (2006), available at http://wdi.umich.edu/files/publications/workingpapers/wp826.pdf (status: 21.02.2013).

Bertero, Elisabetta, and Laura Rondi, "Financial Pressure and the Behaviour of Public Enterprises under Soft and Hard Budget Constraints: Evidence from Italian Panel Data," in: *Journal of Public Economics*, 75 (2000), 73–98.

Black, Alexander J., and Hew R. Dundas, "Joint Operating Agreements: An International Comparison from Petroleum Law," in: *Journal of Natural Resources & Environmental Law*, 8 (1992), 49–80.

Bortolotti, Bernardo, and Mara Faccio, "Government Control of Privatized Firms," in: *Review of Financial Studies*, 22 (2009), 2907–2939.

Boscheck, Ralf, "The Governance of Oil Supply: An Institutional Perspective on NOC Control and the Questions It Poses," in: *International Journal of Energy Sector Management*, 1 (2007), 366–389.

Boycko, Maxim, Andrei Shleifer, and Robert W. Vishny, "A Theory of Privatisation," in: *The Economic Journal*, 106/435 (1996), 309–319.

Bremmer, Ian, and Robert Johnston, "The Rise and Fall of Resource Nationalism," in: *Survival*, 51 (2009), 149–158.

Bressand, Albert, "Foreign Direct Investment in the Oil and Gas Sector: Recent Trends and Strategic Drivers," in: *Yearbook on International Investment Law & Policy 2008–2009* (Oxford University Press, New York, 2009), 117–214.

Bridgman, Benjamin, Victor Gomes, and Arilton Teixeira, "Threatening to Increase Productivity: Evidence from Brazil's Oil Industry," in: *World Development*, 39 (2011), 1372–1385.

Brunnschweiler, Christa, and Simone Valente, "International Partnerships, Foreign Control and Income Levels: Theory and Evidence," in: *CER-ETH Economics Working Paper Series*, 11/154 (2011), available at http://ideas.repec.org/p/eth/wpswif/11-154.html (status: 22.02.2013).

Buchanan, James M., "The Constitution of Economic Policy," in: *The American Economic Review*, 77 (1987), 243–250.

Claessens, Stijn, Simeon Djankov, and Larry HP Lang, "The Separation of Ownership and Control in East Asian Corporations," in: *Journal of Financial Economics*, 58 (2000), 81–112.

Collier, Paul *et al.*, *The National Resource Charter*, November 2010, available at http://naturalresourcecharter.org/(status: February 27, 2013).

Cordesman, Anthony H., *Saudi Arabia Enters the Twenty-first Century: The Political, Foreign Policy, Economic, and Energy Dimensions* (Greenwood Publishing Group, Westport, CT, 2003).

Dincer, Ibrahim, and Bandar Al-Rashed, "Energy Analysis of Saudi Arabia," in: *International Journal of Energy Research*, 26 (2002), 263–278.

Dixit, Avinash K., *The Making of Economic Policy: A Transaction Cost Politics Perspective* (MIT Press, Cambridge, MA, 1998).

Domjan, Paul, and Matt Stone, "A Comparative Study of Resource Nationalism in Russia and Kazakhstan 2004–2008," in: *Europe–Asia Studies*, 62 (2010), 35–62.

Dunning, Thad, and Grigore Pop-Eleches, "From Transplants to Hybrids: Exploring Institutional Pathways to Growth," in: *Studies in Comparative International Development (SCID)*, 38 (2004), 3–29.

Ehrlich, Isaac, Georges Gallais-Hamonno, Zhiqiang Liu, and Randall Lutter, "Productivity Growth and Firm Ownership: An Analytical and Empirical Investigation," in: *Journal of Political Economy*, 102 (1994), 1006–1038.

Eller, Stacy L., Peter R. Hartley, and Kenneth B. Medlock, "Empirical Evidence on the Operational Efficiency of National Oil Companies," in: *Empirical Economics*, 40 (2011), 623–643.

Eweje, Gabriel, "Multinational Oil Companies' CSR Initiatives in Nigeria: The Scepticism of Stakeholders in Host Communities," in: *Managerial Law*, 49 (2007), 218–235.

—, "Environmental Costs and Responsibilities Resulting from Oil Exploitation in Developing Countries: The Case of the Niger Delta of Nigeria," in: *Journal of Business Ethics*, 69 (2010), 27–56.

Galdon-Sanchez, Jose E., and James A. Schmitz, "Competitive Pressure and Labor Productivity: World Iron-Ore Markets in the 1980's," in: *American Economic Review*, 92/4 (2002), 1222–1235.

Gochenour, Thomas D., "The Coming Capacity Shortfall: The Constraints on OPEC's Investment in Spare Capacity Expansion," in: *Journal of Energy Policy*, 20 (1992), 973–982.

Goolsbee, Austan, and Chad Syverson, "How Do Incumbents Respond to the Threat of Entry? Evidence from the Major Airlines," in: *The Quarterly Journal of Economics*, 123 (2008), 1611–1633.

Grayson, Leslie E., *National Oil Companies* (Wiley, Hoboken, NJ, 1981).

Grosfeld, Irenna, and Thierry Tressel, "Competition and Ownership Structure: Substitutes or Complements," in: *Economics of Transition*, 10 (2002), 525–551.

Hanson, Philip, "The Resistible Rise of State Control in the Russian Oil Industry," in: *Eurasian Geography and Economics*, 50 (2009), 14–27.

Hartshorn, Jack E., *Oil Trade: Politics and Prospects* (Cambridge University Press, Cambridge, 1993).

Heinrich, Andreas, "Under the Kremlin's Thumb: Does Increased State Control in the Russian Gas Sector Endanger European Energy Security?," in: *Europe–Asia Studies*, 60 (2008), 1539–1574.

Herb, Michael, "A Nation of Bureaucrats: Political Participation and Economic Diversification in Kuwait and the United Arab Emirates," in: *International Journal of Middle East Studies*, 41 (2009), 375–395.

Janosz, William W., Julia Kou, and Debora L. Spar, "White Nights and Polar Lights: Investing in the Russian Oil industry," in: *Harvard Business Review Case Studies*, June 29, 1995.

Jensen, Michael C., "Organization Theory & Methodology," in: *Accounting Review*, Vol. LVIII/2 (1983), 319–339.

Johnston, Daniel, *International Exploration Economics, Risk, and Contract Analysis* (PennWell Books, Tulsa, OK, 2003).

Jones Luong, Pauline, and Erika Weinthal, *Oil Is Not a Curse: Ownership Structure and Institutions in Soviet Successor States* (Cambridge University Press, New York, 2010).

Khanna, Neha, "On the Economics of Non-Renewable Resources," in: *Binghamton University Department of Economics Working Papers*, 102 (2001), available at www2.binghamton.edu/economics/research/working-papers/pdfs/wp01/WP0102.pdf (status: February 22, 2013).

Klare, Michael T., *Blood and Oil: The Dangers and Consequences of America's Growing Dependency on Imported Petroleum* (Holt Paperbacks, New York, 2004).

Lahn, Glada, Valerie Marcel, John Mitchell, Keith Myers, and Paul Stevens, *Report on Good Governance of the National Petroleum Sector* (Chatham House, Royal Institute of Foreign Affairs, 2007).

Van der Linde, Coby, "The Art of Managing Energy Security Risks," *EIB Papers*, 12/1 (2007), 50–78.

Lippman, Thomas W., *Inside the Mirage: America's Fragile Partnership with Saudi Arabia* (Basic Books, New York, 2004).

Locatelli, Catherine, "The Russian Oil Industry Between Public and Private Governance: Obstacles to International Oil Companies' Investment Strategies," in: *Journal of Energy Policy*, 34 (2006), 1075–1085.

Luciani, Giacomo, "Allocation vs. Production States: A Theoretical Framework," in: *The Arab State* (University of California Press, Berkeley, CA, 1990), 65–84.

Mahdavi, Paasha, *State Ownership and the Resource Curse: A New Dataset on Nationalizations in the Oil Industry*, SSRN 1916590 (2011), available at http://papers.ssrn.com/sol3/papers.cfm?abstract_id=1916590 (status: February 22, 2013).

Marcel, Valerie, *Oil Titans: National Oil Companies in the Middle East* (Brookings Institution Press, Washington DC, 2006).

Mehrara, Mohsen, "Energy Consumption and Economic Growth: The Case of Oil Exporting Countries," in: *Journal of Energy Policy*, 35 (2007), 2939–2945.

Mitchell, Timothy, "Carbon Democracy," in: *Economy and Society*, 38 (2009), 399–432.

—, *Carbon Democracy: Political Power in the Age of Oil* (Verso, New York, 2011).

Mommer, Bernard, *Global Oil and the Nation State* (Oxford University Press, Oxford, 2002).

Myers Jaffe, Amy, and Jareer Elass, "Saudi Aramco: National Flagship with Global Responsibilities," *The James A. Baker III Institute for Public Policy—Rice University Working Papers*, 2007, available at www.bakerinstitute.org/programs/energy-forum/publications/docs/NOCs/Papers/NOC_SaudiAramco_Jaffe-Elass-revised.pdf (status: February 22, 2013).

Nakhle, Carole, "Petroleum Fiscal Regimes: Evolution and Challenges," in: *The Taxation of Petroleum and Minerals: Principles, Problems and Practice* (Routledge, London, 2010), 89–121.

Nickell, Stephen J., "Competition and Corporate Performance," in: *Journal of Political Economy*, 104/4 (1996), 724–746.

Nore, Petter, and Terisa Turner, *Oil and Class Struggle* (Zed Press, London, 1980).

North, Douglass C., "Institutions," in: *Journal of Economic Perspectives*, 5/1 (1991), 97–112.

Persson, Torsten, and Guido Tabellini, *Political Economics: Explaining Economic Policy* (MIT Press, Cambridge, MA, 2002).

PFC Energy, *Forces Shaping Strategic Themes and Hubs*, Presentation in Washington D.C., February 10–11, 2011.

Pinto, Brian, and Sweder van Wijnbergen, "Ownership and Corporate Control in Poland: Why State Firms Defied the Odds," in: *CEPR Discussion Papers*, 1995, available at http://ideas.repec.org/p/cpr/ceprdp/1273.html (status: February 22, 2013).

Pollitt, Michael G., *Ownership and Performance in Electric Utilities: The International Evidence on Privatization and Efficiency* (Oxford University Press, Oxford, 1995).

La Porta, Rafael, Florencio Lopez de Silanes, Andrei Shleifer, and Robert W. Vishny, "Law and Finance," in: *Journal of Political Economy*, 106/6 (1998), 1113–1155.

Roland, Gerard, "Understanding Institutional Change: Fast-Moving and Slow-Moving Institutions," in: *Studies in Comparative International Development (SCID)*, 38 (2004), 109–131.

Ross, Michael, *The Oil Curse: How Petroleum Wealth Shapes the Development of Nations* (Princeton University Press, Princeton, NJ, 2012).

Saghi, Gabor, "What Is Economic Policy For?" in: *Periodica Polytechnica: Social and Management Sciences*, 10/1–2 (2002), 191–200.

Schlager, Edella, and Elinor Ostrom, "Property-Rights Regimes and Natural Resources: A Conceptual Analysis," in: *Land Economics*, 68/3 (1992), 249–262.

Schmitz, James A., "What Determines Productivity? Lessons from the Dramatic Recovery of the US and Canadian Iron Ore Industries Following Their Early 1980s Crisis," in: *Journal of Political Economy*, 113 (2005), 582–625.

Schmitz, James A., and Arilton Teixeira, "Privatization's Impact on Private Productivity: The Case of Brazilian Iron Ore," in: *Review of Economic Dynamics*, 11 (2008), 745–760.

Shapiro, Carl, and Robert D. Willig, "Economic Rationales for Privatization in Industrial and Developing Countries," in: *The Political Economy of Public Sector Reform* (Westview Press, Boulder, CO, 1990), 22–54.

Shleifer, Andrei, and Robert W. Vishny, "A Survey of Corporate Governance," in: *The Journal of Finance*, 52/2 (1997), 737–783.

Stevens, Paul, "Oil Markets," in: *Oxford Review of Economic Policy*, 21 (2005), 19–42.

—, "National Oil Companies and International Oil Companies in the Middle East: Under the Shadow of Government and the Resource Nationalism Cycle," in: *The Journal of World Energy Law & Business*, 1 (2008), 5–30.

Stevens, Paul, and Evelyn Dietsche, "Resource Curse: An Analysis of Causes, Experiences and Possible Ways Forward," in: *Journal of Energy Policy*, 36 (2008), 56–65.

Thurber, Mark C., David R. Hults, and Patrick R. P. Heller, *The Limits of Institutional Design in Oil Sector Governance: Exporting the "Norwegian Model"*, Presentation at ISA Annual Convention, 2010, available at http://iis-db.stanford.edu/pubs/22836/Thurber_Hults_and_Heller_ISA2010_paper_14Feb10.pdf (status: February 22, 2013).

—, "Exporting the 'Norwegian Model': The Effect of Administrative Design on Oil Sector Performance," in: *Journal of Energy Policy*, 39 (2011), 5366–5378.

Tietenberg, Thomas H., and Lynne Lewis, *Environmental and Natural Resource Economics* (HarperCollins Publishers, New York, 2000).

Tinbergen, Jan, *On the Theory of Economic Policy* (North-Holland Publishers, Amsterdam, 1952).

Tordo, Silvana, *Fiscal Systems for Hydrocarbons: Design Issues* (World Bank Publications, Washington DC, 2007).

—, "Countries' Experience with the Allocation of Petroleum Exploration and Production Rights: Strategies and Design Issues," in: *World Bank Working Papers*, 2009, available at https://openknowledge.worldbank.org/handle/10986/5954 (status: February 22, 2013).

Tordo, Silvana, Brandon S. Tracy, and Noora Arfaa, *National Oil Companies and Value Creation* (World Bank Publications, Washington DC, 2011).

United Nations General Assembly, *Resolution 1803 (XVII) of 14 December 1962 "Permanent Sovereignty over Natural Resources"*, available at http://untreaty.un.org/cod/avl/ha/ga_1803/ga_1803.html (status: 27.02.2013).

United States (U.S.) Government Accountability Office (GAO), *Oil and Gas Royalties. A Comparison of the Share of Revenue Received from Oil and Gas Production by the Federal Government and Other Resource Owners*, May 2007, available at www.gao.gov/products/GAO-07-676R (status: February 27, 2013).

—, *Oil and Gas Royalties. The Federal System for Collecting Oil and Gas Revenues Needs Comprehensive Reassessment*, Report to Congressional Requesters, September 2008, available at www.gao.gov/new.items/d08691.pdf (status: February 27, 2013).

Vernon, Raymond, *Sovereignty at Bay: The Multinational Spread of US Enterprises* (Basic Books, New York, 1971).

Victor, Nadejda, "On Measuring the Performance of National Oil Companies (NOCs)," in: *PESD Stanford University Working Papers*, 64 (2007), available at http://pesd.stanford.edu/publications/nocperformance (status: February 22, 2013).

Vitalis, Robert, *America's Kingdom: Mythmaking on the Saudi Oil Frontier* (Stanford University Press, Redwood, CA, 2006).

Vivoda, Vlado, "Resource Nationalism, Bargaining and International Oil Companies: Challenges and Change in the New Millennium," in: *New Political Economy*, 14 (2009), 517–534.

Waelde, Thomas W., "International Energy Investment," in: *Energy Law Journal*, 17 (1996), 191–215.

Wolf, Christian, "Does Ownership Matter? The Performance and Efficiency of State Oil vs. Private Oil (1987–2006)," in: *Journal of Energy Policy*, 37 (2009), 2642–2652.

Wolf, Christian, and Michael G. Pollitt, "The Welfare Implications of Oil Privatisation: A Social Cost-Benefit Analysis of Norway's Statoil," in: *Cambridge Working Papers in Economics*, 907 (2009), available at www.eprg.group.cam.ac.uk/wp-content/uploads/2009/02/binder13.pdf (status: February 22, 2013).

Woodhouse, Erik J., "The Obsolescing Bargain Redux? Foreign Investment in the Electric Power Sector in Developing Countries," in: *Journal of International Law and Politics*, 38 (2006), 121–220.

Zhang, Anming, Yimin Zhang, and Ronald Zhao, "Impact of Ownership and Competition on the Productivity of Chinese Enterprises," in: *Journal of Comparative Economics*, 29 (2001), 327–346.

3 Current readings of Energy Studies and theory development

The field of Energy Studies is dominated by engineers, economists, and only partly, by political scientists. The latter are largely interested in Energy Security, Energy and Conflict, Energy and Resource Curse,[1] and only marginally in Energy Policy, which has been treated as a given for other social, political or economic phenomena. This book refuses to take ownership and control structures endogenously. Rather, it seeks to explain them at the sectoral level—namely, in the oil upstream industry worldwide. The motivation for this endeavor is straightforward: the existing scholarship does not provide a clear set of explanations for the variation in the upstream sector policies pursued by petroleum-rich countries globally. There is a lot of research on the effects of control—both at the firm/company level and at the sectoral level. Yet, apart from three exceptions, each of them with a clear regional focus, there is no research on who controls the oil upstream sector and why so. This is the niche where this book seeks to make a contribution by examining the control structures in the oil upstream industry worldwide.

In Benton's opinion,

> the point of interest of most research to date has not been on how the sector's structure came into existence but on how it shapes other things. Yet, energy policy and thus sector structures varies dramatically across nations and over time. That energy sector structures affect hydrocarbons income and thus politics and economics means that the energy policy choice should be of analytic concern.

(2008, 4f.)

Apart from her exceptional research, the two other bodies of work which address energy policy as the actual object of study belong to Palacios (2001, 2002, and 2003) and Jones Luong and Weinthal (2001, 2006, and 2010). The former—similarly to Benton (2008)—has a regional focus on Latin America while the latter grapples with the post-Soviet space. Still, it should be emphasized that Jones Luong and Weinthal's work is particularly tailored to the Resource Curse literature. Acknowledging the variation in ownership and control structures in the hydrocarbons sector, they seek to establish the missing link between these structures and institutional outcomes.

By comparison with Jones Luong and Weinthal's research, studies from the NOC literature like Eller *et al.* (2011), Hartley and Medlock (2008), Victor (2007) or Wolf (2009) look into the effect of ownership and control structures—and not at the industry level but at the firm level. More explicitly, these studies statistically compare the performance of National Oil Companies (NOCs) with that of International Oil Companies (IOCs). In a more qualitative manner, the other strand in the NOC literature—epitomized by the World Bank studies (2008; Tordo *et al.* 2011), Rice University's Baker Institute Project on NOCs, "The Role of National Oil Companies in International Energy Markets," Victor *et al.* (2012), or Marcel's (2006) research—entails individual case studies and comparative analyses of NOCs and IOCs, where ownership and control structures are again taken for granted.

The literature which comes closest to the focus of the present book is the Nationalization/Expropriation one. However, nationalization is different from control (and ownership) structures in the sense that the former is just one instance in time whereas the latter is the product of an economic policy. In other words, the former is moment-based whereas the latter is medium to long term. The former is an event, for which the onset is registered—hence, a binary variable: 0 or 1. In turn, ownership and control structures take values along a continuous scale running from 0 to 1, or in percentages, from null to 100 percent. Nationalization is just one means to increase control in the oil upstream sector (Kobrin 1984, 329). After the wave of industry nationalizations in the 1970s, nationalization has been pursued only marginally and if so, mostly at the firm level—that is, divestment of the private assets in a single consortium and not in the entire sector. Such an example is Yacimentos Petroliferos Fiscales (YPF) in Argentina (see *The Economist*, April 21, 2012). Being involuntary and most of the time based on coercion, nationalization has a negative effect on the reputation of the country in question, boasts its level of political risk and is likely to discourage foreign investments (Conklin 2002; Hashmi and Guvenli 1992; Makhija 1993; Rios-Morales *et al.* 2009; Vedpuriswar 2002). Increase in state control does not need to occur through nationalization—instead, it may happen through a market-based transaction with satisfactory outcomes for all parties involved and with no negative consequences for the country's image. A case in point is TNK-BP in Russia (*Financial Times*, October 22, 2012).

Despite all the differences discussed above, in the absence of a better-suited scholarship on ownership and control structures in the energy sector, this book draws on the Nationalization/Expropriation studies in order to identify potential determinants of upstream sector policies. Critical thinking will be the guiding criterion in the discrimination between driving forces that may work differently in the case of state control than in the case of Nationalization/Expropriation.

Moreover, the review will encompass a number of studies which do not belong to the Nationalization/Expropriation literature. Some come from Macroeconomics and Comparative Politics and address the effects of markets and political institutions on economic policies. Others are geological and more generally, engineering studies which examine the role of technical and technological factors in the energy sector. They will all be discussed with the aim of providing a comprehensive picture of the broader Energy Studies, which deal with potential determinants of energy policy choices. Based on the reviewed scholarship, the second part of this chapter seeks to bring together the identified explanations into a unitary analytical framework of upstream sector policies in oil producing countries worldwide.

3.1 Review of drivers and drives in the energy sector

A large number of scholars from different disciplines have emphasized that energy policies result from the interaction of economic, technological and politico-institutional factors (Kohl 1982; Lindberg 1977; Lucas and Papaconstantinou 1985; Nolan and Thurber 2010; Prontera 2009; Tordo 2009). Therefore, an interdisciplinary approach is not just in order but rather imperative in addressing the subject. In the search for potential determinants of oil upstream sector policies, this section groups the spate of Energy Studies according to (1) technical, (2) economic and (3) politico-institutional explanations. They will be discussed separately below.

3.1.1 Technical explanations

Two main technical arguments are advanced by the studies reviewed in this section to explain energy policy choices. These are the level of risk as a function of geological conditions and the extent of technical expertise and technological know-how of the state-owned oil company.

Upstream activities—exploration, development and production—are associated with different levels of risk and uncertainty. Risk stands for the probability of occurrence of a discrete event such as the discovery of commercial reserves and is expressed through a single probability estimate (Ross 1997, 2004; Simpson *et al.* 2000). In turn, uncertainty

> reflects the inability to estimate a value exactly, such as the remaining recoverable volumes of oil and/or gas from a producing field (. . .). In contrast with risk, it should be noted that in the case of a continuous distribution of uncertainty, probability values are *always* related to a range of estimates, never to a discrete outcome.
>
> (Ross 2004, 42)

Petroleum scientists and engineers looked into the relevance of these two elements at different stages in an upstream sector project.

In the exploration phase, risks are related to the "chance of success" (Rose 1992). Success in petroleum exploration may take different forms: "geological success" or initial discovery of hydrocarbons, "completion success" when the exploratory well is finalized following estimations that the petroleum accumulations will pay off the costs of development, and "commercial success" when "a discovery well (. . .) finds a field capable of paying for all subsequent costs of development drilling, completion, surface equipment, operation costs, and wellhead taxes, returning a reasonable profit margin plus the cost of an equivalent number of dry holes" (ibid., 71). Thus, the discovery of hydrocarbons does not guarantee production.

In reference to a well-established resource classification system, Ross distinguishes between discovered versus undiscovered resources, and if discovered, commercial versus sub-commercial, as follows: "[u]ndiscovered recoverable volumes (i.e. not penetrated by a well) are classified as prospective resources, discovered sub-commercial volumes as contingent resources and discovered, commercial volumes (. . .) are classified as reserves" (2004, 45). In time, with new technological developments and/or opening of a viable market, contingent resources may become commercially recoverable. Just as well, enhancements in drilling and well-scanning techniques are likely to increase precision in the measurement of recoverable volumes from an oilfield and turn theretofore prospective resources into discovered ones. In this respect, Joseph Stiglitz showed that "[i]n cases where costs of extraction are currently high, and might be lowered over time with the progress of technology, the return to waiting may be higher than on any other investment the government might make" (quoted in Tordo *et al.* 2011, 6).

Both in the exploration phase and (if commercial production starts) at further stages in the project, uncertainty exists and refers to the range of estimated recoverable volumes, quality of oil and reservoir/oilfield characteristics, infrastructure requirements, production plan, and/or operational costs (Suslick and Schiozer 2004, 1).

In the NOC literature, petroleum risks and uncertainties are rather generally defined. They stand for the geological difficulties raised by the identification and if the case, development and production of petroleum deposits (Nolan and Thurber 2012; Tordo *et al.* 2009, 2011). For example, Nolan and Thurber (2012) discuss levels of petroleum risk[2] relative to the lifecycle of the petroleum province. If operations are at the "frontier"—i.e. either in the initial stage of "new province exploration and field development," or in the "tertiary recovery" (final) phase—the challenges are high. Yet, if the operations are in the "secondary recovery" stage—i.e. once hydrocarbons are discovered, wells completed and commercial reserves estimated—the level of petroleum risk is low and also, the challenges of extraction are rather low.

Oilfields geology leads to different costs of exploration and production. More specifically,

> [o]il at great depth or under deep water costs more to drill and operate the well than shallow, dry-land oil deposits. Extracting cost per barrel is lower for larger fields because of fixed investment. In the United States, for example, the direct operating cost for primary oil recovery for a 12000-foot well is 3 times more expensive than that of a 2000-foot one. [. . .] It is estimated that the cost of producing oil can range from as little as $2 per barrel in the Middle East to more than $15 per barrel in some fields in the United States.
>
> (Tsui 2006, 7)

In this sense, Al-Attar and Alomair (2005) find that exploration and production costs explain the type of legal and fiscal arrangements adopted by oil producing countries. These upstream costs are determined by geological conditions, as follows: high costs are caused either by mature oilfields with declining production capacities, which demand improved oil recovery (IOR) technologies, or by low onshore reserves, which force high-risk operations in deep waters offshore. The scholars show that in the past few decades following the era of nationalizations, low cost countries (with costs of $4.65 per barrel or less) like Kuwait, Iran, Venezuela or Saudi Arabia have adopted service agreements. Medium cost countries (with costs between $4.65 and $8.50 per barrel) like Kazakhstan, Oman, the United Arab Emirates (UAE), Nigeria or Indonesia have used production-sharing agreements (engaging NOCs and IOCs) and partly, royalty-tax systems. The case of devolving full control to IOCs through the royalty-tax systems is also found in high cost countries (with costs of more than $8.50 per barrel) like Norway. The conclusion is that oil producing countries opened their upstream and agreed to grant control rights to IOCs only in order to share risks and attract both capital and technology for the development of difficult oilfields and the exploration of new frontiers. This finding is reiterated by Tordo *et al.*'s study (2011, 4) with respect to gas reserves.

Varied geological conditions further create different requirements as concerns technology and technical expertise. Across the NOC literature, technological and technical requirements stand for the capabilities which are necessary to run the exploration, development and production activities at a given level of geological risk (Nolan and Thurber 2012; Tordo *et al.* 2011). They are generally assessed in relation to the technical know-how and technological facilities of the oil producing country in question, or its economic capacity to acquire these (Bressand 2009; Nolan and Thurber 2012; Stevens 2008).

On this note, it is widely agreed that "technological gaps play a fundamental role in the rise of foreign-control regimes or international partnerships" (Brunnschweiler and Valente 2011, 11). More specifically, countries operating at the frontier—i.e. either with new discoveries of natural resources or in the mature phase—need advanced technology to jumpstart or respectively, uphold production. In the absence of such technology, and also expertise to use it if acquired, oil producing countries are willing to open up the upstream sector to foreign firms (Benton 2008, 3; Mahdavi 2011, 7).

In an analysis of energy policy choices in Latin American oil-endowed countries, Palacios considers that one of the main explanations for the variance therein is the operational performance, financial viability and corporate culture of the NOC. It is yet not clear how these characteristics of the NOC directly impact the openness/closeness of the energy sector. By the power of example, the scholar claims that the liberalization measures in Argentina in the 1980s were triggered by the financial and operational deficiencies of the YPF whereas in Brazil and Venezuela they were the product of the relative efficiency and profitability of their internationally-exposed NOCs, Petrobras and respectively, PDVSA. The argument becomes fuzzier with the Mexican example, where despite the NOC's operational inefficiencies foreign participation was barred as "Mexico's Pemex developed into an inward-oriented institution with a powerful clientelist structure that opposes any liberalization of the sector" (Palacios 2002, 4).

Nonetheless, from an empirical perspective, the comparative insights from Marcel on Kuwait and Abu Dhabi (2006, 146), the discussion of Kazakhstan and KMG EP's capacity from Tordo *et al.* (2011, II: 19) and more generally, the case studies from Victor *et al.* (2012) and Rice University's Project on NOCs showcase how the lack of technological capacity and technical prowess in some NOCs has paved the way to cooperation with private oil companies in several oil producing countries.

All in all, geological conditions and, related to these, the requisites for technical expertise and technological prowess, are the take-away from the reviewed studies of this section in terms of potential determinants of oil upstream sector policies worldwide.

3.1.2 Economic explanations

The Nationalization/Expropriation literature advances two main economic explanations for the variation in energy sector policies—namely, oil prices and the resource dependence of the producer country. Regarding oil prices, the general argument poses that higher oil prices may lead to large wind-falls, which create stronger incentives for countries to nationalize their upstream sector (Iwayemi and Skriner 1986; Shiravi and Ebrahimi 2006; Warshaw 2012).

Quite interestingly, Duncan (2006) advances alternative hypotheses for the impact of output resource price on the behavior of oil producing countries. On the one side, the "opportunistic explanation" posits that in so-called "boom years" higher real prices of the mineral resource lead to higher probability of expropriation. On the other side, the "desperation explanation" contends that in "bust years" it is exactly low real prices which increase the probability of expropriation.[3] Duncan's statistical testing finds evidence for the opportunistic interpretation, but none whatsoever for the alternative.

Similarly, Guriev *et al.* (2011) find that expropriations in the oil industry are significantly more likely to occur when oil prices are high for longer periods. However, the scholars show that while high oil prices might create incentives to nationalize the relatively more valuable resources, "given the costs of nationalization, it is not immediately clear why a government would respond to a positive oil price shock with nationalization rather than simply imposing higher taxes" (ibid., 303). In their opinion, since nationalizations are value-reducing, it is the quality of institutions which allow (or not) for inefficient decisions to be made.[4]

In her study of energy policy choices in several Latin American countries (i.e. Venezuela, Bolivia, Brazil, Colombia, Ecuador, and Mexico), Benton finds that high commodity prices create emboldening conditions for host governments to "increase state intervention without significant capital flight" (2008, 11). Correspondingly, low prices combined with unfavorable geological conditions, such as high production costs and declining hydrocarbons reserves, make governments more likely to give control to private oil companies.

Using the same estimate of "oil price shocks" like in Guriev *et al.*'s model (2011, 312), which captures not the absolute oil price but the deviation of the oil price from its long-term trend, Warshaw (2012) also detects a positive relation between oil price shocks and expropriations, yet not Christensen (2011) who comes across insignificant results. While the latter admits the fact that above normal prices may trigger expropriations, he cautions on the possible reverse causality which cannot be controlled for in the absence of good instruments. As to the effect of low prices on the likelihood of privatizations, Warshaw (2012) is unable to run a multivariate regression due to insufficient data. In this regard, Benton affirms that "[p]eriods of falling state hydrocarbons revenues (. . .) lead governments from all ideological positions to support a role for private investment in the hydrocarbons sector" (2008, 3) and provides empirical support for this in her comparative case study design. Also, the statistical analysis of expropriation and privatization cycles conducted by Chang *et al.* (2010) renders expropriations more likely in high-price regimes and privatizations more prone to occur when prices fall under a certain threshold. By and large, higher oil prices are thus associated with an increased propensity to nationalize the hydrocarbons sector in oil producing countries.

Another economic explanation of the variation in energy sector policies—encountered in the NOC literature—is oil dependence. This refers to the level of "rentierism" in the domestic economy of the oil-rich country, which measures the degree of national economic dependency on the petroleum revenues. In reference to the Rentier State[5] studies from the Resource Curse literature, "a country is typically considered a rentier when rent revenues make up a minimum percentage of all government revenues, when oil exports make up the bulk of GDP, or something similar" (Herb 2009, 376).

NOC studies identify a negative correlation between oil dependence and upstream sector openness. For example, the World Bank study by Tordo *et al.* (2011) on the performance and mandate of NOCs finds that the higher the oil dependence is, the more favorable the macro-fiscal and sectoral policy to the NOC becomes, which translates into more state control in the upstream sector. Palacios also uncovers the same relation by assuming that oil exporters are more reliant on oil revenues than self-sufficient countries or oil importers. As of the early 2000s in Latin America, Argentina, Peru, Bolivia and Brazil—as oil importers and energy self-sufficient countries— are indicative of relatively more open oil industries than oil exporters such as Mexico, Colombia, Ecuador and Venezuela (Palacios 2001, 2002).

In "a state-centered and revenue-centered approach, focusing on the incentives that resource wealth may pose to incumbent political elites," Dunning (2005, 452) models the political economy of a resource-rich country in order to evaluate the political causes and consequences of resource dependence. Based on the empirical evidence provided by the cases of post-independence Botswana, Mobutu's Zaire, and Suharto's Indonesia, the analysis shows that resource dependence is in fact nurtured by state elites: "resource dependence is the outcome of strategic decisions by incumbent elites to limit the extent to which political opponents can challenge their power" (ibid., 475). More explicitly, political elites, who are in control of the main source of revenues in most resource-rich countries— that is, from the natural resource sector—conscientiously and rationally uphold high economic reliance on natural resource revenues because this way, the would-be political opposition remains financially dependent on the state or in other words, on the distribution of economic and political patronage from the current ruling elites. The scholar theorizes that "fiscal crisis and economic contraction do not cause regime change or political instability in resource-dependent states, because promoting resource dependence is itself a way that elites can block the viability of challenges to incumbent power" (ibid.). While Mobutu's Zaire is an epitome of this logic, the same rationale can be applied to other countries from Sub-Saharan Africa (e.g. Gabon under Omar Bongo, Zambia under Kenneth Kaunda) or the Persian Gulf (e.g. Saudi Arabia, Kuwait).

Finally, Jones Luong and Weinthal (2010) factor alternative revenues—as a reverse way of defining oil dependence—into their explanations of ownership and control structures in post-Soviet energy producer countries. The difference to Dunning's study is that Jones Luong and Weinthal allow for variance in the economic structure of the petroleum-rich states as they deem that ruling elites may be able to draw revenues from alternative sources—and not only from the hydrocarbons sector—in order to reward their base of supporters and appease or repress their potential contenders. Nonetheless, in the absence of other productive economic sectors, ruling elites prefer direct access to the source of income for the sake of their survival in power. In short, higher oil dependence is conducive to more state ownership and control in the energy sector. The way in which institutions may play a role in mediating this relation and restrain the behavior of state leaders (despite high oil dependence) will be explored next.

3.1.3 Institutional explanations

In the following, the institutions which may influence the policy choice in the energy sector, more generally and in the oil upstream sector, more particularly, are identified. For this purpose, works originating from the Nationalization/Expropriation scholarship, Comparative Political Studies, International Political Economy and Public Policy Studies are considered. While scholars agree upon the high relevance of domestic institutions for the decisions in the energy sector, the role of international institutions remains controversial. Still, for the sake of comprehensiveness, both domestic and international institutional factors are addressed below.

The scholarship on the impact of institutions on the energy sector has mostly dealt with formal constraints. Due to measurement and data collection issues, informal constraints have been difficult to grasp and widely seen as undergirding the formal institutional configurations or in other words, Roland's (2004) "fast-moving institutions."[6]

3.1.3.1 Domestic institutions

Before looking into the existing scholarship for ways in which domestic institutions may shape decisions in the energy sector, the preferences of different stakeholders have to be addressed. Given the limited research on energy policy choices, insights into the preferences of stakeholders should contribute to a better understanding of policy outcomes.

3.1.3.1.1 PREFERENCES IN THE DOMESTIC ARENA

Most studies in the Nationalization/Expropriation literature are state-centric and set off from the premise that "[s]tate leaders make strategic decisions in the petroleum industry in order to enhance their ability to stay in power. Moreover, these decisions are likely to be structured by leaders' political incentives (i.e., retaining power) and constraints" (Warshaw 2012, 36). With survival in power as the main goal of their actions, state leaders in petroleum-rich countries may take different paths to attain this.

A few scholars have contended that in the special case of these countries, the best means for state leaders to achieve their key objective of staying in power is "sovereignty maximization" in the resource sector (Jones Luong and Weinthal 2001, 2006, 2010; Ross 1999; Wantchekon 1999). More explicitly, "state leaders in energy-rich states will choose development strategies that enable them to achieve a maximum level of sovereignty over their natural resources without thus threatening their continued rule"—in short, "all prefer more rather than less sovereignty, which translates into more rather than less control over their natural resources" (Jones Luong and Weinthal 2001, 373f.). This way, state leaders and more broadly, ruling elites can secure their political survival by transferring wealth to their acolytes and bribe-taking officials (Click and Weiner 2010, 800), buying-off opposition (Jones Luong and Weinthal 2001, 2010; Kalyuzhnova and Nygaard 2008), funneling investment into preferential projects (Guriev *et al.* 2009, 2011; Kobrin 1985; Li 2009; Peltzman 1989) or boosting up their own bank accounts for dark times. Also, they can benefit the society more directly by providing public goods, investing in capital accumulation and long-run industrial development, or by pursuing welfare programmes (Click and Weiner 2010; Horn 1995; McPherson 2003; Stevens 2003; van den Bosch 2012).

Several comparative political studies develop formal models to show how different strategies pursued by state leaders and ruling elites in order to retain power may harm the economic and societal structures in the medium to long run. These strategies are thus economically irrational yet politically rational. In an excellent article, Garrett and Lange prove that state leaders cannot think of the societal good in the long term "lest they lose power in the interim. The simplest way to avoid this fate is to distribute benefits to the groups whose support brought them to office—even if this has significant costs for macroeconomic performance" (1995, 631). Though not economically efficient, such a distributional game may prove politically efficient for the state leaders and ruling elites both in democracies and autocracies. While in democracy, redistribution may take more refined forms than in weakly institutionalized polities and include the larger public through provision of public goods (van den Bosch 2012), the logic behind remains the same. Analyses of kleptocratic and more broadly, authoritarian regimes bring to light a spate of other strategies—besides redistribution for patronage—which state leaders and ruling elites apply in order to remain in power (e.g. "divide-and-rule" in Acemoglu *et al.* 2003, "politics of fear" in Miquel 2007, obstruction of technological and institutional changes in Acemoglu and Robinson 2006, steady pursuit of a resource-dependent economy in Dunning 2005). These strategies seem more likely to be followed when the national economy is fuelled by rents and ruling elites have a firm grip on these rent flows (Acemoglu and Robinson 2006; Dunning 2005).

Still, the behavior of state leaders and ruling elites is constrained to a greater or lesser degree by domestic institutional arrangements. Taking this one step further, economic policies may be "the product of the unconstrained choice of decision-makers" or may be explicated by "structurally determined factors such as (. . .) the existence of particular types of state institutions"

(Karl 1997, 8). In other words, the economic policy for the natural resource sector may be the exclusive choice of the state leader and ruling elites or instead, may be the product of several institutional actors with varied preferences. Furthermore, the policy choice may or may not take account of the public preference.

On this note, there is a clearly discernible preference among people in resource-rich countries with regard to the natural resources. This is for state ownership and state control. In this regard, Marcel noted the discrepancy between public opinion and national industry in terms of the favored policy for energy resources:

> [i]n most of the producing countries, there is a gap between the national industry's concerns and those of society and its opinion leaders. The national media, popular opinion and other political institutions, such as the parliament, are often fiercer in resisting foreign involvement.
>
> (2006, 39)

A public survey conducted in 18 Latin American countries in 2008 inquired about the industries which "should be mostly in the hands of the State and which should mostly be in the hands of private companies" (Latinobarometro 2008). The results are self-explanatory: 80 percent of the respondents consider that the oil industry should be run by the state and its oil company, and not by private firms.

There might be several reasons for this public preference concerning the energy resources. As noted under the motives for NOC creation (section 2.2.3.1), one of these may have historical roots and go back to the colonial times when resource-endowed countries could neither decide over nor benefit from the exploits of the national wealth. The international oil majors made the most of the opportunities and kept both operations and profits in their hands. The nationalizations in the 1970s inaugurated a new relation of power between public and private actors in the energy sector and made the state and ideally thereby, the people in the host countries the prime beneficiaries of the fruit born by the energy sector.

Notably, this penchant for state ownership and control of the natural resources is not unique to publics in developing countries. A study of opposition to the privatization of the energy sector finds that:

> countries with recent campaigns [against private involvement, *my note*] include high-income countries like France, Germany, and the USA; transition countries such as Hungary and Poland; middle-income countries such as Mexico, South Africa, and Thailand; and low-income countries like Ghana, Honduras, and India. (. . .) The opposition is clearly not limited to factors that are peculiar to developing countries.
>
> (Hall *et al.* 2005, 292)

The widespread belief across the public is that "even if privatization enhances efficiency, the bulk of its benefits accrue to a privileged few—shareholders, managers, domestic or foreign business interests, those connected to the political elite—while the costs are borne by the many" (Kikeri and Nellis 2002, 2). In turn, state ownership and state control are more likely to protect the right of the people to the fair share from their national wealth. In an analysis of nationalization and privatization cycles, Chang *et al.* find evidence for the fact that nationalization is more likely to occur under conditions of rampant or worsening inequality at the national level and "especially when the rents from natural resource or utility companies are perceived as benefitting only a minority" (2010, 7). In this sense, private ownership and control of the resource sector were seen as conducive to a more inequality-plagued society in Latin America and Southeast Asia throughout the twentieth century, with the benefitting minority composed of "external foreigners" (i.e. "Western imperialists") and some "foreigners within" (defined by ethnicity and social class/cast affiliation) (Chua 1995). The recent history of Latin America indicates that the public continues to vie for a hydrocarbon sector run by the state. Examining resource nationalism and leftism in Latin American producer countries in the late 1990s and early 2000s, Benton finds that "[e]ven where governments have not reversed energy sector strategies, politicians that advocate increased state presence in the sector have gained considerable support among the citizens" (2008, 1).

More generally, the issue of sharing proceeds to the benefit of the population turns particularly sensitive for strategic industries, among which the resource sector/extractive industry is just one example (Boubakri *et al.* 2009; Kobrin 1984). Just the last few years have witnessed the creation of several institutions worldwide, which are aimed at increasing protectionism of the strategic subsoil assets, more narrowly and the strategic industries, more broadly. Such examples are: the U.S. Department of Treasury's Committee on Foreign Investment in the United States (CFIUS); Italian Law Decree No. 21 of March 15, 2012 (all cited in Portolano 2012) introducing new veto powers for the government in strategic sectors such as defense and national security, energy, transport and communications; Mongolia's "Law on the Regulation of Foreign Investment in Business Entities Operating in Sectors of Strategic Importance"; the Strategic Foreign Investment Law of May 17, 2012 (Portolano 2012); Russia's Strategic Sectors Law (Heath 2009; Pomeranz 2010) etc.

Given the main goal of the state leader and ruling elites, on the one hand, and the preferences of the public concerning the resource sector, on the other hand, the question which needs to be further explored is how the policy choice is made given the domestic institutional setup.

3.1.3.1.2 DOMESTIC INSTITUTIONAL CONSTRAINTS

In the Nationalization/Expropriation literature, scholars disagree about the effect of institutions on the probability of nationalization. When institutions are boiled down to executive constraints (as defined and operationalized by Marshall and Jaggers 2006), different results are obtained. Guriev *et al.* (2009, 2011) find support for the conventional wisdom that nationalizations are more likely to occur in countries with fewer executive constraints. Reproducing Guriev *et al.*'s statistical analysis, Warshaw (2012) confirms the results and puts forward an 8.4 percent probability of nationalization for countries with the lowest level of executive constraints in the Polity IV dataset as compared to 1.6 percent probability for the countries with the highest score on executive constraints. However, the addition of a time counter to the model, as a way to account for time trends across countries, nullifies the effect of executive constraints on the probability of expropriation. The computation of the marginal effect of executive constraints over time is instructive about its contingency on time. More explicitly, until the mid-1970s, countries with low levels of executive constraints were more prone to expropriate than those with high levels of executive constraints, but the relation does not hold thereafter (Warshaw 2012, 51f.).

Contrary to Guriev *et al.* (2009, 2011) and Warshaw (2012), Duncan (2006) uses the overall polity score—that is, the combined index for the level of democracy and autocracy of the political institutions by country per year (Marshall and Jaggers 2006)—and concludes that democracies are in fact more likely to expropriate than dictatorships because the leaders have a short-term horizon (compared to autocrats) and therefore tend to be more price sensitive. While Duncan examines the relation between institutions and expropriations on a database of major exporters of bauxite, copper, lead, nickel, silver, tin and zinc, Guriev *et al.* (2009, 2011) and Warshaw (2012) study nationalizations in the oil industry. Nonetheless, the contradictory results cannot be explained by the different types of minerals analysed by the scholars, as Mahdavi's (2011) research proves. In an innovative approach to nationalizations in the oil industry, Mahdavi supports Guriev *et al.*'s claim that democracies—this time, operationalized with the Polity score (similarly to Duncan)—are less inclined to nationalize their oil sector (Mahdavi 2011, 19).

In the face of these inconsistent findings, Christensen (2011) proposes a more nuanced take on political institutions in relation to nationalizations by distinguishing between limitations on executive power and the degree of political competition for office. For the operationalization of the former, the academic uses the polcon index from Henisz (2000). Drawing on Duncan's (2006) database, Christensen shows that "checks and balances that constrain the executive power always reduce the probability of expropriations" (2011, 91). In turn, competition for political power, measured by Vanhanen's (2014) index, has a non-linear effect: at higher levels of political competition, the probability of expropriation is very low because the president needs the support of the people to stay in power and is forced to think of the long-term interest of the country; yet, for low to medium levels of political competition, the probability of expropriation is high since the short-term (financial) gains are favored over long-term costs.

In an insightful study of economic policy choices, van den Bosch (2012) explains that these should be seen as a function of two regime characteristics which define the distribution of power at the polity level—namely, the level of political inclusiveness and the time horizon of the regime. Focusing on patterns of rent allocation, the scholar finds that predatory behavior on the side of the state leader can be restrained in highly inclusive polities. When a large share of the population is involved in the political process, incumbent elites cannot choose an economic policy favorable to their rule only, but instead need to take note of the people's preferences. In his research on resource nationalism in Latin America, Mares (2010) validates this relation. As for cross-generational time horizons, these—in combination with low political constraints—are prone to make the incumbents take control of a larger share of the economy "in order to structure economic sectors in a way that is beneficial to the existing political order" (van den Bosch 2012, 10). Short-term horizons are likely to turn incumbent elites into myopic actors exclusively interested in maximizing rent extraction—yet, institutional constraints fortunately play a moderating role in this case.

Moving now to the few exceptions, which seek to explain economic policy choices in the energy sector, it should be emphasized that one of these three bodies of work does not address the role of institutions. Instead, Luisa Palacios' PhD dissertation (2001) and further articles (2002, 2003) propose two alternative explanations, already professed in the sections above. These are: the position of an oil producing country in the international energy market, which is in fact a measure of the national oil dependence, and the NOC's performance and corporate culture, which can be explained by the historical context of the oil industry in the country. Palacios focuses on variation in energy policies in Latin America until 2000.

Benton (2008) further preserves the research focus both content- and region-wise, but concentrates on the time period ranging from the late 1990s to 2008. The scholar advances three factors to account for variation in energy policy across Latin American countries: first, the ideological preference of governments compared to the current energy policy, second, executive–legislative dynamics, and third, state fiscal revenues. The revenues work as an antecedent variable in the model and are a result of prices, extraction costs, production levels, and depletion ratios. Similarly to van den Bosch's (2012) theoretization of how economic policy choices are shaped, Benton considers that governments think in terms of their own costs and benefits, hold firm energy policy preferences—this time, not necessarily with respect to the time horizon (like in van den Bosch 2012) but to ideology (left/right leaning)—and are constrained in their choices by the legislature. In a case study analysis of six Latin American countries, i.e. Venezuela, Bolivia, Brazil, Colombia, Ecuador, and Mexico, which span the full range of upstream energy policy variance, the scholar draws the following conclusions: given high fiscal revenues, left-leaning governments wish to increase the state presence in hydrocarbons production—however, their success "depends on whether they can build the support necessary to pass them into law in national legislatures" (Benton 2008, 36). Under conditions of low revenues from hydrocarbons production, left-leaning governments are willing to engage with private oil companies and make the latter undertake investment risks—so left-leaning governments are then prone to favor liberalizing measures of the production sector provided that they raise sufficient legislative support. By comparison, centrist or right-leaning governments prefer minimal or no state presence in the energy sector (to the extent possible) and are likely to increase the government take through royalties and taxes when profits from hydrocarbons production are high (such as when prices are high).

The third body of research examining upstream energy policies belongs to Pauline Jones Luong and Erika Weinthal (2001, 2006, and 2010). While both Palacios (2001, 2002, and 2003) and Benton (2008) propose explanations for Latin American countries, the regional focus shifts in case of Jones Luong and Weinthal to the post-Soviet states. Also, while the first two bodies of work examine variance in energy policies at a given point in time, Jones Luong and Weinthal address initial energy development strategies adopted by Soviet successor countries once they gained independence and obtained authority over their own energy reserves or first discovered them. This difference in content is relevant because provided that there is already an alternative functioning economic sector in place, state leaders may opt for postponing hydrocarbons development and production for a later stage. In short, the structure of costs and incentives based on which state leaders and ruling elites make their preferred choice may function differently in the case of initial energy development strategies. This is at least the conclusion to be gathered by comparing Jones Luong and Weinthal's outcomes with those provided by the other two scholars. Also, this may be due to the fact that in contrast to Benton (2008) or more generally, van den Bosch (2012), Jones Luong and Weinthal do not recognize the fact that other preferences might exist concerning the energy policy besides that of the state leader/ruling elites and that the elites may be affected by these other preferences in their policy-making under given circumstances.

The model which runs across Jones Luong and Weinthal's work for a decade (2001, 2006, and 2010) starts off from the assumption that state leaders not only wish to stay in power but also like to retain as much control as possible over hydrocarbons production. The examination of the initial formation of energy development strategies in five post-Soviet countries (i.e. Turkmenistan, Uzbekistan, Azerbaijan, Russia and Kazakhstan) bring the scholars to the conclusion that state leaders choose strategies so that they provide them with sufficient resources on the one hand, to sustain the cleavage structure that offered their main base of support and on the other hand, to placate or overpower rival cleavages that (might) pose a challenge to their rule. The interaction between two key variables—availability of alternative revenues and the level of distributional conflict or synonymously, political contestation[7] (2001, 394; 2010, 301)—helps explain how energy development choices are made. Based on the relation between available resources and costs incurred by political contestation (e.g. patronage, financial concessions to or suppression/oppression of rivals), state leaders may embark on their most preferred energy development strategy, which is state ownership with state control, or slide down their rank of preferences (2006, 256). The more intense the contestation of the existing ruling order is, the more resources the rulers need to survive in power (2006, 256 ff.; 2010, 303). The most preferred energy development policy can be adopted only under conditions of high alternative revenues and low political contestation. At the other extreme, when alternative resources are low and distributional conflict is high, rulers opt for private foreign ownership because of "domestic pressures to generate revenue immediately so as to both maintain status quo support and diffuse a potent challenge to their continued rule" (2006, 258). In the absence of both political contestation and alternative exports, state leaders grant control rights to foreign oil companies but retain the ownership over the energy resources—this way, the energy sector is likely to be rapidly developed and able to generate proceeds in a short time. In turn, when both political contestation and alternative revenues are high, states leaders transfer both ownership and control of resources to domestic private oil companies. The explanation provided in this case is that thereby, state leaders "can bolster dominant patronage networks and appease the emerging rival one" (2006, 259).

The statistical analysis in Jones Luong and Weinthal (2010) in Chapter 9, "Taking domestic politics seriously: explaining the structure of ownership over mineral resources," draws on an original dataset of petroleum-rich developing countries worldwide from the late 1800s to 2005 and finds support for the conceptual model presented above. Regarding the dependent variable in their analysis, this captures "de jure" ownership and control structures given that the coding is based on "their respective constitutions, official laws and regulations governing the mineral sector, and (where available) mineral contracts between the state and corporate entities (foreign and domestic) operating in the petroleum sector" (2010, 311). Yet the Ordinary Least Squares (OLS) regression applied with a categorical four-value dependent variable and the use of the same operationalization for two different arguments—namely, oil price as a proxy for both international market conditions, and technology and difficulty of extraction—make the results rather disputable from a statistical point of view.

Notably, while Benton recognizes the contribution of Jones Luong and Weinthal's work to the existing scholarship, she expresses skepticism about the applicability of their model to the Latin American petroleum-rich countries, where "fiscal resources and politics appear to work differently" (2008, 5). The same skepticism could be raised when trying to apply the results to the Middle East region.

To wrap up, this section reviewed the existing scholarship on the role of domestic institutions for energy policy choices. To this end, it accounted both for the preferences of domestic actors and for the constraints set by specific institutional arrangements on the policy choice in the energy sector. The views vary largely as to the effect of political institutions. There might be several reasons for this: first, varied understandings of political institutions and ways of operationalization; second, relatively different objects of study: nationalization/expropriation versus economic policy/energy policy; third, varied geographical foci: global analysis in most Nationalization/Expropriation studies versus regional focus (e.g. Latin America, post-Soviet region) in the few studies on energy policies. Given the diversity of the lessons drawn from this discussion of domestic institutional arguments, substantial thought needs to be put into the construction of the analytical framework—but before that, the role of international institutions has to be reviewed.

3.1.3.2 International institutions

Research on the effect of international institutions on energy policies is scarce, to say the least. Despite its size, the scholarship fails to agree about the leverage of international institutions over energy policy choices in oil producing countries. One camp considers that there is a clear effect mostly under the form of policy diffusion. Yet another camp denies any impact of international rules and international organizations at the domestic level.

So if there is an effect of international organization on the energy sector, pundits claim that this is the result of policy diffusion (Guler *et al.* 2002; Henisz *et al.* 2005; Jones Luong and Weinthal 2010; Polillo and Guillén 2005) and not of global pressures for policy convergence like in other sectors (Brune *et al.* 2004; Frank *et al.* 2000; Kaufmann *et al.* 2007; Polak 1997). The most relevant international organization for petroleum-rich countries is the Organization of Petroleum-Exporting Countries (OPEC), but there is a large spate of international and regional institutions which may have an indirect impact on domestic energy policies—such as the UN and its agencies, International Energy Agency, Energy Charter Conference and Secretariat etc. (Florini and Sovacool 2009; Waelde 2004).

In their statistical analysis of ownership and control structures in petroleum-rich countries, Jones Luong and Weinthal (2010) control for both an international and a regional diffusion effect. The former is aimed at capturing the adoption of an initial development strategy before or after the foundation of OPEC in 1960, the latter is accounted for through dummy variables for Middle East and North African (MENA) and respectively, Latin America. The effect of international policy diffusion is statistically significant, yet of a very low magnitude. Arguably, this finding supports the idea that OPEC encouraged developing countries to nationalize their oil resources and also, buoyed up the power of petroleum exporting countries to the detriment of foreign oil companies. This argument is corroborated by other scholars and practitioners like Chaudhry (1989, 1994, 1997), Mahdavi (2011), Morse (1999), Vernon (1971). In turn, the regional demonstration effect as in policy diffusion generated by regional organizations or alternatively, by a sense of regional belonging does not pass the significance tests in Jones Luong and Weinthal's statistical chapter (2010, 318). Based on a dataset of 62 oil producing countries spanning the 1905 to 2005 period, Mahdavi (2011) identifies a strong effect of OPEC membership on nationalizations from the early 1960s to the early 1980s, and none whatsoever afterwards.

Notably, the other camp departs from energy-related international and regional organizations and discusses the effect of broader international institutions on the behavior of oil producing countries. Voeten and Ross (2011) contend that due to their position in the global economy, these countries afford the luxury of remaining politically unglobalized—that is, of not taking note of international organizations. Due to the relevance of their resources on international markets, these states may engage in international trade and develop economic relationships with other countries, but have the autonomy to shape domestic policies at their own will. In brief, "interdependence does not necessarily lead to greater political integration: domestic politics plays a critical role" (ibid., 10).

Jensen and Johnston's (2011) research further shows that governments in resource dependent countries are less sensitive to reputation costs than other administrations. Despite evidence of heightened political risk, these countries do not encounter difficulties in attracting foreign direct investment (FDI). The reason behind this is that for most investors the unusually high benefits from extraction outweigh the expropriation risks. Therefore, resource-rich countries dare to expropriate more and thereby, to disregard property rights and bilateral international treaties with foreign investors.

In an analysis of global energy governance, Goldthau (2012) concludes that due the large number of energy-related institutions and the lack of coordination between them, at present there are no effective mechanisms to influence energy policies at national level.

Given the limited research and disparate views on the impact of international institutions on oil producing countries and their policy choices, it is yet unclear how and to what extent OPEC or any other international or regional institution might have an influence on the upstream sector policy at national level. This needs to be further unpacked.

In conclusion, this chapter sought to review all major studies which look into potential drivers of energy policy choices. Political Science studies were supplemented with macroeconomic analyses, geological and more broadly, engineering articles in order to unlock all possible explanations. At this point, it should be clear that there is no systematic research on the determinants of upstream sector policies in oil producing countries worldwide. While global studies of oil nationalizations such as Guriev *et al.* (2009, 2011), Mahdavi (2011) or Warshaw (2012) build a valuable starting point for this book, their focus remains still different from the one envisaged here—namely, oil upstream sector policy. The scholarly works which grapple with explanatory models of energy policy choices adopt, in turn, specific regional foci such as Latin America and the post-Soviet space. With the exception of Jones Luong and Weinthal's (2010) statistical analysis, there is no work that explains how ownership and control structures in the energy upstream sector are forged globally.

On these premises, this book tries to build an analytical framework which addresses upstream sector policies across the population of oil producing countries worldwide. To this end, the potential driving forces identified in the existing literature—of technical, economic and institutional nature—are first reunited in a comprehensive scheme and then tested in the framework of a mixed-methods design which includes not just a large statistical analysis but also case studies from the Middle East region.

3.2 The analytical framework—explaining oil upstream sector policies worldwide

In the pursuit of explanations for the variation in upstream sector policies across oil producing countries worldwide, this framework takes a state-centered approach, similarly to Dunning (2005), Jones Luong and Weinthal (2001, 2006, and 2010), van den Bosch (2012) and others. State leaders and/or the collective executive are at the epicenter of policy-making in the petroleum sector.[8]

In the context of a theory-building exercise, the three types of explanations reviewed in section 3.1 (i.e. technical, economic, and institutional) are brought together to explicate how upstream sector policy choices worldwide are made. For a graphical representation of the analytical framework, see Figure 3.1.

The context is defined by three exogenous elements, which means that they cannot be influenced in the short run or if at all, only marginally, by the state leader and/or the executive in the oil producing countries. These are: the two main factors identified in the review of technical arguments (geological conditions and NOC capabilities) and the economic driver set on the international markets (oil price). The context creates the basis for decision-making and should inform economically efficient policies in the upstream. Nonetheless, the state leader and the executive are primarily interested in political survival. That is why they are likely to make policy choices which are economically inefficient or less efficient as long as they serve their main goal of remaining in power. Political rationality takes precedence over economic rationality. Yet these policy options may be strongly constrained by two key domestic forces: by the country's economic dependence on oil revenues (oil dependence) and the limits on executive

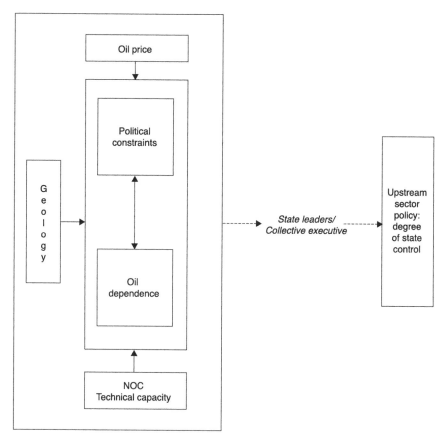

Figure 3.1 Analytical framework—explaining oil upstream sector policies worldwide

power (executive constraints). The former is the other economic driver pinned down in the reviewed literature while the latter belongs to the institutional arguments. The way each of these elements influences the upstream sector policy choice in the oil industry worldwide is spelled out in the next sections.

3.2.1 The context

The premises for policy-making are created on the one hand, by geology and the technical capacity of the NOC, and on the other hand, by the international oil price. Following the layout of a SWOT analysis, the domestic elements (i.e. geology and NOC technical capacity) can be seen as strengths or weaknesses, whereas the international factor (i.e. oil price) may represent an opportunity or a threat for the executive and the state leader of an oil producing country in their policy choice for the upstream sector.[9]

Depending on the geology in place (uncertainty and risks, resource endowment, complexity of extraction), an oil-rich country may require more or less advanced technology to produce the commercially successful oilfields. Geology is an exogenous factor because it works as a given for the state leaders and the collective executive.

The other domestic factor of contextual nature is the NOC with its level of technical know-how and operational capacity. State leaders can influence the technological assets of an NOC by financing the purchase of cutting-edge technology and can also enhance the operational capacity by rendering more autonomy to the NOC. However, if not present, the technical capacity and expertise of the staff to deliver on operations and be able to run processes—all the more in the presence of complex technology—require time to be built up. Attaining operational performance in any company, but even more so in a state-owned enterprise, is usually not a short-term task. This is why, domestically, at a given point in time, state leaders are faced with a given state-owned operator in the upstream which may be more or less technically and technologically capable of meeting the geological conditions on the ground.

The contextual picture is further shaped by the oil price level which is set on international energy markets and which may create an opportunity or a threat to producer governments. More specifically, when the oil price is high, the executive/state leaders may perceive an opportunity to fill the coffers by buttressing control over the upstream. Instead, low oil prices are likely to build a threat for budgetary distress in oil producing countries, especially when the largest chunk of government revenues comes from oil exports.

3.2.2 Domestic constraints

Given the contextual parameters exposed above, the executive and/or the state leader are free or less so to pursue their preferred policy in the upstream sector depending on two main domestic constraints: the level of political constraints and the extent of the country's dependence on oil revenues. These represent the decisive factors for the policy adopted in the upstream given that they shape the state leader and the executive's actual room for maneuver. The preference of the state leader and the executive concerning the upstream sector policy is the product of their primary goal of survival in power and not of some political ideology. In short, it is a product of political rationality.

The contextual factors may or may not be seriously taken into account as the state leader and the executive might willingly opt for highly distortionary upstream sector policies for the economy at large (i.e. economically irrational), yet highly effective at dismantling opposition (i.e. politically rational). In other words, the upstream sector might be run inefficiently and unsustainably when considering the geological characteristics of the oil production base, the technical and technological requirements for the oilfields to be productive long term and not least, the economic considerations (depletion policy and performance of NOCs vs. IOCs) raised by the international oil price. However, this freedom for the executive and state leader is conferred or barred by the political institutions and the economic situation of the country. Thus, political constraints and oil dependence build the explanatory variables of main interest in this analytical framework and are considered to have the largest explanatory power.

In the following section, the effect of each of these factors will be discussed. A set of hypotheses will emerge to account for the variation in upstream sector policies, with the weight being placed on domestic economic and politico-institutional factors.

3.2.3 Hypotheses

The first hypothesis addresses the effect of geology.

> H1: *The more complex the geological conditions are, the less state control is expected in the upstream sector (or in other words, the more control for IOCs).*

Geological conditions refer to the level of technical complexity associated with the extraction process. The logic behind this hypothesis is that when oilfields are difficult to produce because of the geological characteristics in place (geography such as onshore/offshore, stage of production cycle, particularities of the minerals etc.), the state leader and the executive in oil producing countries are more likely to open up the upstream sector, diversify the risks and attract foreign investors by giving up upstream control rights.

The same situation is likely to occur when the NOC is lacking the technical capacity to explore and produce, given the geological conditions. Out of necessity, state leaders and host governments will have to be more lenient towards foreign oil companies and grant control rights in the upstream sector as they need their know-how and expertise. Consequently,

> H2: *The lower the technical capabilities of the NOC are, the less state control is expected in the upstream sector (and thus, the more control rights given up to IOCs).*

So, if the requirements for technical expertise and technological know-how created by the geological conditions are high—due to, for example, mature oilfields, high depletion rate, low reserve additions—and the oil producing country through its state-owned oil company does not dispose of the necessary capabilities to explore and produce, the upstream operations need to be allocated to private companies under favorable contractual terms for the latter. Complex geological conditions and low technical capacity of the

NOC creates a less favorable context for the state leaders and the executive to design the upstream sector policy.

Internationally, oil markets can set a price for the barrel of crude oil which might tempt state leaders to increase control in the upstream sector. The existing Expropriation/Nationalization scholarship widely agrees about the proclivity to nationalize—and by extension here, increase state control— with higher oil prices. More explicitly, since higher oil prices promise more funds and potentially, large windfalls, state leaders are presented with strong incentives to increase the state control in the upstream sector or even take over completely. Therefore,

> *H3: The higher the oil prices are, the more state control is expected in the upstream sector.*

Moving on from the contextual factors, the economic conditions of a country may incur serious constraints on the choices made by state leader and their executive. High oil reliance of an oil-rich country makes the executive and state leader more dependent on oil revenues and entices them into managing the resources directly—that is, through direct state control. This way, they can be sure of sufficient financial means to cover government expenses and if the case, defeat opposition and marshal public support through distribution of patronage or public goods, for example.

Notably, the revenues or synonymously, the source of national income can stem from the oil sector or alternative sources such as other productive economic sectors or foreign aid. Therefore, based on the degree of national reliance on oil revenues or synonymously, on the extent of "rentierism," state leaders might need and opt (or not) for more upstream control.

In short,

> *H4: The higher the oil reliance of the oil producing state is, the more state control is expected in the upstream sector.*

The focus will now turn to domestic political constraints. Due to the controversies about the impact of political institutions on the probability of nationalization in the reviewed literature, the essentials need to be first uncovered. For this purpose, the least common denominator for the oil producing countries (irrespective of the polity) is taken at the core of this hypothesis, i.e. executive constraints. These cover:

> the extent of institutional constraints on the decision-making powers of the chief executive, whether an individual or a collective executive. Such limitations may be imposed by any "accountability groups". In Western democracies these are usually legislatures. Other kinds of accountability groups are the ruling party in a one-party state; councils of nobles or powerful advisors in monarchies; the military in coup-prone polities;

and in many states a strong, independent judiciary. The concern is therefore with the checks and balances between the various parts of the decision-making process.

(Marshall *et al.* 2010, 64)

Based on the differences between nationalization/expropriation and increase in state control (outlined above), this book considers that the effect of political/executive constraints works differently in the case of industry control. More specifically, if there are any checks and balances in the oil producing country, these would not restrain the increase in state control, as with nationalization (cf. most of the Nationalization/Expropriation studies), but rather trigger it. The hypothesis posits that:

H5: The higher the level of executive constraints is, the more state control is expected in the upstream sector.

The logic behind this is that if there exists any accountability group in the country, then this would represent the preference of the public. In reference to academic research and public surveys (reviewed in section 3.1.3.1.1), the people in oil producing countries and by inference, the parliament (where existent) as the voice of the people opt for more state control over the oil sector and the national resources, which are considered a "national patrimony." In case the collective executive and/or the state leader might prefer opening up the upstream sector to IOCs—due to financial considerations related to their stay in power as in a quicker generation of revenues through IOCs to uphold their basis of support and mollify the contenders—the accountability group would reduce the feasibility of policy change.

Furthermore, while two of the three bodies of work explaining energy policy choices (Benton 2008 and Jones Luong and Weinthal 2001, 2006, and 2010) contend that the interaction between oil dependence and political constraints is crucial in order to understand the policy outcome, it is not clear how this interaction shapes energy policy choices. As discussed and quoted in section 3.1.3.1.2, Benton (2008) dismisses Jones Luong and Weinthal's model (2006) as not applicable in the Latin American region. In fact, the findings of their bodies of research (Benton versus Jones Luong and Weinthal) are contradictory. This is why this interaction cannot and will not be hypothesized here but instead exploratorily tested in the statistical analysis. That will uncover whether the effect of executive constraints is indeed moderated by the extent of oil dependence, and if so, how.

While arguably still rough, the analytical framework proposed here tries to be a simple and rather straightforward approach to oil upstream sector policies worldwide. This is why factors like international institutions, whose relevance for domestic energy economic policies has been seriously disputed, are not incorporated. Instead, they will be exploratorily included as controls into the statistical analysis. Moreover, since this framework seeks to be applicable to oil producing countries globally, the least common denominator for political constraints has been taken. This is the extent of executive constraints, or synonymously, the independence of the executive authority to make decisions. Conflating measures like polity/regime quality or alternatively, more sophisticated classifications of political institutions may only obscure the relation to energy policies, as the disagreements in the existing literature show. Last but not least, the construction of this analytical framework has sought to take note of the different preferences and goals existent in the domestic arena, and not just of those pertaining to the ruling elites, by contrast to Jones, Luong, and Weinthal's (2001, 2006, and 2010) model, for example. The analytical framework will be set under test in the empirical analysis following in the next two chapters. Based on the statistical findings and the comparative insights provided by the two case studies in Chapters 4 and 5, respectively, the framework will be revised and enhanced.

Notes

1 These three broad themes are clearly overlapping as, for example, conflict can be seen as a materialization of the Resource Curse or could be subsumed to Energy Security. However, Energy Studies can be arguably regarded as encompassing these three main lines of research plus Energy Policy.
2 These scholars muddle the concepts of uncertainty and risk as they subsume the former to the latter: "[p]etroleum investment risk has two distinct components: the uncertainty of an outcome and the capital invested in this outcome" (Nolan and Thurber 2010, 5). Yet the take-away from their study regarding the challenges encountered over the lifecycle of an oilfield/petroleum province is insightful.
3 The rationale is framed in terms of state consumption and utility gains, starting off from two assumptions: first, the long-run steady state for the national economy with FDIs implies a higher level of consumption than in the no-FDI steady state; second, while expropriation raises domestic income in the short run, this falls back to the no-FDI steady state level in the long run. On these premises, opportunistic countries expropriate in price "boom years" when the return on capital is high and their utility gain from expropriation is expected to be the highest (compared to future periods). In turn, desperate countries with very low consumption are likely to expropriate in "bust years" because they are less willing to substitute consumption inter-temporally and their marginal utility from expropriation would be very high in the short term.
4 For this, please see section 3.1.3.1.2 on politico-institutional explanations above.
5 For a detailed discussion of this, please see Mahdavy (1970), Beblawi and Luciani (1987), and Luciani (1990).
6 Another frequent distinction made by scholars of institutionalism concerns the relation between institutions and organizations. The latter build "purposive entities designed by their creators to maximize wealth, income, or other objectives defined by the opportunities afforded by the institutional structure of the society" (North 1990, 73). As Simons and Martin put this, "[w]hile we recognize the distinction between institutions and organizations, many of our arguments apply to both. In the actual practice of research, the distinction

between institutions and organizations is usually of secondary importance" (2002, 194). The same is the case in this book.
7 This is measured by investigating whether "(a) there exists a cleavage structure that could work as a viable alternative to the current basis for dispensing patronage, (b) political parties and/or social movements based on such an alternative cleavage have emerged and gained popular support, and (c) these parties and movements have in fact made demands for greater resources" (Jones Luong and Weinthal 2001, 374).
8 For this, please see discussion in section 2.1.1, "The institutional set-up for policy-making in the oil sector".
9 This approach follows the rationale of a SWOT analysis, which is a method used to assess the strengths, weaknesses, opportunities and threats involved in a project of any kind. A SWOT analysis can be conducted for a product, place, industry or person. In short, it seeks to map out the internal factors (i.e. strengths and weaknesses) and external factors (i.e. opportunities and threats) that are favorable and unfavorable for attaining a specific objective for, e.g., a company or an industry (Hill and Westbrook 1997).

References

Acemoglu, Daron, and James A. Robinson, "Economic Backwardness in Political Perspective," in: *American Political Science Review*, 100 (2006), 115–131.

Acemoglu, Daron, James A. Robinson, and Thierry Verdier, "Kleptocracy and Divide-and-Rule: A Model of Personal Rule," in: *NBER Working Papers*, 10136 (2003), available at www.nber.org/papers/w10136 (status: February 22, 2013).

Al-Attar, Abdulaziz, and Osamah Alomair, "Evaluation of Upstream Petroleum Agreements and Exploration and Production Costs," in: *OPEC Review*, 29 (2005), 243–266.

Beblawi, Hazem, and Giacomo Luciani, *The Rentier State* (Routledge Kegan & Paul, London, 1987).

Benton, Allyson, "Political Institutions, Hydrocarbons Resources, and Economic Policy Divergence in Latin America," in: *2008 APSA Annual Meeting Papers*, available at www.international.ucla.edu/economichistory/summerhill/benton1.pdf (status: February 21, 2013).

van den Bosch, Marie Aliénor, "Natural Resources, Economic Performance and the Politics of Public Finance in Rentier States," in: *Princeton University Working Papers*, 2012, available at www.princeton.edu/politics/graduate/courses/seminars/comparative-politics-semi/MAvdB_POL591Paper_MAY4.pdf (status: February 21, 2013).

Boubakri, Narjess, Jean-Claude Cosset, and Omrane Guedhami, "From State to Private Ownership: Issues from Strategic Industries," in: *Journal of Banking & Finance*, 33 (2009), 367–379.

Bressand, Albert, "Foreign Direct Investment in the Oil and Gas Sector: Recent Trends and Strategic Drivers," in: *Yearbook on International Investment Law & Policy 2008–2009* (Oxford University Press, Oxford, 2009), 117–214.

Brune, Nancy Elizabeth, Geoffrey Garrett, and Bruce Kogut, "The International Monetary Fund and the Global Spread of Privatization," in: *IMF Papers*, 51/2 (2004).

Brunnschweiler, Christa, and Simone Valente, "International Partnerships, Foreign Control and Income Levels: Theory and Evidence," in: *CER-ETH Economics*

Working Paper Series, 11/154 (2011), available at http://ideas.repec.org/p/eth/wpswif/11-154.html (status: February 22, 2013).

Chang, Roberto, Constantino Hevia, and Norman Loayza, "Privatization and Nationalization Cycles," in: *World Bank Policy Research Working Paper Series*, 5029 (2010), available at http://ideas.repec.org/p/wbk/wbrwps/5029.html (status: February 22, 2013).

Chaudhry, Kiren Aziz, "The Price of Wealth: Business and State in Labor Remittance and Oil Economies," in: *International Organization*, 43 (1989), 101–145.

—, "Economic Liberalization and the Lineages of the Rentier State," in: *Comparative Politics*, 27/1 (1994), 1–25.

—, *The Price of Wealth: Economies and Institutions in the Middle East* (Cornell University Press, Ithica, NY, 1997).

Christensen, Jonas Gade, "Democracy and Expropriations," in: *University of Bergen, Department of Economics Working Papers in Economics*, 6 (2011), available at www.uib.no/filearchive/wp06.11.pdf (status: February 22, 2013).

Chua, Amy L., "Privatization-Nationalization Cycle: The Link between Markets and Ethnicity in Developing Countries," in: *Yale Law School Faculty Scholarship Series*, 342 (1995), available at http://digitalcommons.law.yale.edu/fss_papers/342/ (status: February 22, 2013).

Click, Reid W., and Robert J. Weiner, "Resource Nationalism Meets the Market: Political Risk and the Value of Petroleum Reserves," in: *Journal of International Business Studies*, 41 (2010), 783–803.

Conklin, David W., "Analyzing and Managing Country Risks," in: *Ivey Business Journal*, 66 (2002), 36–56.

Duncan, Roderick, "Price or Politics? An Investigation of the Causes of Expropriation," in: *Australian Journal of Agricultural and Resource Economics*, 50 (2006), 85–101.

Dunning, Thad, "Resource Dependence, Economic Performance, and Political Stability," in: *Journal of Conflict Resolution*, 49 (2005), 451–482.

The Economist, *Argentina's energy industry: Fill 'er up*, April 21, 2012, available at www.economist.com/node/21553070 (status: February 27, 2013).

Eller, Stacy L., Peter R. Hartley, and Kenneth B. Medlock, "Empirical Evidence on the Operational Efficiency of National Oil Companies," in: *Empirical Economics*, 40 (2011), 623–643.

Financial Times, *Rosneft to pay $55bn in TNK-BP takeover*, co-authored by Guy Chazan and Catherine Belton, October 22, 2012, available at www.ft.com/intl/cms/s/0/9776a77a-1c39-11e2-a14a-00144feabdc0.html#axzz2M5hnMRYg (status: February 27, 2013).

Florini, Ann, and Benjamin K. Sovacool, "Who Governs Energy? The Challenges Facing Global Energy Governance," in: *Journal of Energy Policy*, 37 (2009), 5239–5248.

Frank, David John, Ann Hironaka, and Evan Schofer, "Environmental Protection as a Global Institution," in: *American Sociological Review*, 65 (2000), 122–127.

Garrett, Geoffrey, and Peter Lange, "Internationalization, Institutions, and Political Change," in: *International Organization*, 49 (1995), 627–655.

Goldthau, Andreas, "A Public Policy Perspective on Global Energy Security," in: *International Studies Perspectives*, 13 (2012), 65–84.

Guler, Isin, Mauro F. Guillén, and John Muir Macpherson, "Global Competition, Institutions, and the Diffusion of Organizational Practices: The International

Spread of ISO 9000 Quality Certificates," in: *Administrative Science Quarterly*, 47 (2002), 207–232.

Guriev, Sergei, Anton Kolotilin, and Konstantin Sonin, *Determinants of Nationalization in the Oil Sector: A Theory and Evidence from Panel Data*, Working Paper, 2009, available at www.dsg.fohmics.net/Portals/Pdfs/Kolotilin.pdf (status: February 22, 2013).

—, "Determinants of Nationalization in the Oil Sector: A Theory and Evidence from Panel Data," in: *Journal of Law, Economics, and Organization*, 27 (2011), 301–323.

Hall, David, Emanuele Lobina, and Robin De La Motte, "Public Resistance to Privatisation in Water and Energy," in: *Development in Practice*, 15 (2005), 286–301.

Hartley, Peter, and Kenneth B. Medlock III, "A Model of the Operation and Development of a National Oil Company," in: *Energy Economics*, 30 (2008), 2459–2485.

Hashmi, M. Anaam, and Turgut Guvenli, "Importance of Political Risk Assessment Function in US Multinational Corporations," in: *Global Finance Journal*, 3 (1992), 137–144.

Heath, Jesse, "Strategic Protectionism? National Security and Foreign Investment in the Russian Federation," in: *George Washington International Law Review*, 41 (2009), 465–501.

Henisz, Witold J., "The Institutional Environment for Economic Growth," in: *Economics & Politics*, 12/1 (2000), 1–31.

Henisz, Witold J., Bennet A. Zelner, and Mauro F. Guillén, "The Worldwide Diffusion of Market-oriented Infrastructure Reform, 1977–1999," in: *American Sociological Review*, 70 (2005), 871–897.

Herb, Michael, "A Nation of Bureaucrats: Political Participation and Economic Diversification in Kuwait and the United Arab Emirates," in: *International Journal of Middle East Studies*, 41 (2009), 375–395.

Hill, Terry, and Roy Westbrook, "SWOT Analysis: It's Time for a Product Recall," in: *Long Range Planning*, 30/1 (1997), 46–52.

Horn, Murray J., *The Political Economy of Public Administration: Institutional Choice in the Public Sector* (Cambridge University Press, Cambridge, 1995).

Iwayemi, Akin, and Edith Skriner, "Determinants of Upstream Investment in the Oil Industry: An Empirical Analysis," in: *OPEC Review*, 10 (1986), 335–344.

Jensen, Nathan M., and Noel P. Johnston, "Political Risk, Reputation, and the Resource Curse," in: *Comparative Political Studies*, 44 (2011), 662–688.

Jones Luong, Pauline, and Erika Weinthal, "Prelude to the Resource Curse Explaining Oil and Gas Development Strategies in the Soviet Successor States and Beyond," in: *Comparative Political Studies*, 34 (2001), 367–399.

—, "Rethinking the Resource Curse: Ownership Structure, Institutional Capacity, and Domestic Constraints," in: *Annual Review of Political Science*, 9 (2006), 241–263.

—, *Oil Is Not a Curse: Ownership Structure and Institutions in Soviet Successor States* (Cambridge University Press, Cambridge, 2010).

Kalyuzhnova, Yelena, and Christian Nygaard, "State Governance Evolution in Resource-Rich Transition Economies: An Application to Russia and Kazakhstan," in: *Journal of Energy Policy*, 36 (2008), 1829–1842.

Karl, Terry Lynn, *The Paradox of Plenty: Oil Booms and Petro-States* (University of California Press, Berkeley, CA, 1997).

Kaufmann, Daniel, Aart Kraay, and Massimo Mastruzzi, *The Worldwide Governance Indicators Project: Answering the Critics* (World Bank, World Bank Institute, Global Programs, and Development Research Group, Growth and Macroeconomics Team, 2007).

Kikeri, Sunita, and John Nellis, "Privatization in Competitive Sectors: The Record to Date," *World Bank Policy Research Working Paper*, ISSN: 1813-9450, 2002, available at http://elibrary.worldbank.org/content/workingpaper/10.1596/1813-9450-2860 (status: February 22, 2013).

Kobrin, Stephen J., "Expropriation as an Attempt to Control Foreign Firms in LDCs: Trends from 1960 to 1979," in: *International Studies Quarterly*, 28/3 (1984), 329–348.

—, "Diffusion as an Explanation of Oil Nationalization or the Domino Effect Rides Again," in: *Journal of Conflict Resolution*, 29 (1985), 3–32.

Kohl, Wilfrid L., *After the Second Oil Crisis: Energy Policies in Europe, America, and Japan* (Lexington Books, 1982).

Latinobarometro – Opinion Publica Latinoamericana, *Latinobarometro Report 2008*, available at www.latinobarometro.org/latino/latinobarometro.jsp (status: February 27, 2013).

Li, Quan, "Democracy, Autocracy, and Expropriation of Foreign Direct Investment," in: *Comparative Political Studies*, 42 (2009), 1098–1127.

Lindberg, Leon N., *The Energy Syndrome: Comparing National Responses to the Energy Crisis* (Lexington Books, Lanham, MD, 1977).

Lucas, Nigel, and Dimitri Papaconstantinou, *Western European Energy Policies: A Comparative Study of the Influence of Institutional Structure on Technical Change* (Oxford University Press, Oxford, 1985).

Luciani, Giacomo, "Allocation vs. Production States: A Theoretical Framework," in: *The Arab State* (University of California Press, Berkeley, CA, 1990), 65–84.

Mahdavi, Paasha, *State Ownership and the Resource Curse: A New Dataset on Nationalizations in the Oil Industry*, SSRN 1916590 (2011), available at http://papers.ssrn.com/sol3/papers.cfm?abstract_id=1916590 (status: February 22, 2013).

Mahdavy, Hossein, "The Patterns and Problems of Economic Development in Rentier States: The Case of Iran," in: *Studies in the Economic History of the Middle East* (Oxford University Press, Oxford, 1970), 428–467.

Makhija, Mona Verma, "Government Intervention in the Venezuelan Petroleum Industry: An Empirical Investigation of Political Risk," in: *Journal of International Business Studies*, 24/3 (1993), 531–555.

Marcel, Valerie, *Oil Titans: National Oil Companies in the Middle East* (Brookings Institution Press, Washington DC, 2006).

Mares, David R., "Resource Nationalism and Energy Security in Latin America: Implication for Global Oil Supplies," *The James A. Baker III Institute for Public Policy – Rice University Working Papers*, 2010, available at http://bakerinstitute.org/publications/EF-pub-MaresResourceNationalismWorkPaper-012010.pdf (status: February 22, 2013).

Marshall, Monty G., and Keith Jaggers, *Polity IV Country Reports*, 2006, available at www.systemicpeace.org/polity/keynew.htm (status: 22.02.2013).

Marshall, Monty, Keith Jaggers, and Ted Gurr, *Polity IV Country Reports*, 2010, available at www.systemicpeace.org/polity/polity06.htm (status: February 22, 2013).

McPherson, Charles, "National Oil Companies: Evolution, Issues, Outlook," in: *Fiscal Policy Formulation and Implementation in Oil-Producing Countries* (IMF Publications, Washington DC, 2003), 184–203.

Miquel, Gerard Padró I., "The Control of Politicians in Divided Societies: The Politics of Fear," in: *The Review of Economic Studies*, 74 (2007), 1259–1274.

Morse, Edward L, "A New Political Economy of Oil?" in: *Journal of International Affairs*, 53/1 (1999), 1–29.

Nolan, A. Peter, and Mark C. Thurber, "On the State's Choice of Oil Company: Risk Management and the Frontier of the Petroleum Industry," in: *PESD Stanford Working Papers*, 99 (2010), available at http://iis-db.stanford.edu/pubs/23057/WP_99,_Nolan_Thurber,_Risk_and_the_Oil_Industry,_10_December_2010.pdf (status: February 22, 2013).

—, "On the State's Choice of Oil Company: Risk Management and the Frontier of the Petroleum Industry," in: *Oil and Governance. State-Owned Enterprises and the World Energy Supply* (Cambridge University Press, New York, 2012), 121–170.

North, Douglass C., *Institutions, Institutional Change and Economic Performance* (Cambridge University Press, Cambridge, 1990).

Palacios, Luisa, *Explaining Policy Choice in the Oil Industry: A Look at Rentier Institutions in Mexico and Venezuela (1988–1999)*, PhD Thesis, Johns Hopkins University, 2001.

—, "The Petroleum Sector in Latin America: Reforming the Crown Jewels," in: *SciencePo CERI Studies*, 88 (2002), available at www.sciencespo.fr/ceri/en/content/petroleum-sector-latin-america-reforming-crown-jewels (status: February 22, 2013).

—, "An Update on the Reform Process in the Oil and Gas Sector in Latin America," in: *Japan Bank for International Cooperation (JBIC) Working Papers*, September 2003.

Peltzman, Sam, "The Control and Performance of State-Owned Enterprises: Comment," in: *Privatisation and State-Owned Enterprises: Lessons from the United States, Great Britain and Canada* (Kluwer, Boston, 1989), 69–76.

Polak, Jacques, "The World Bank and the IMF: A Changing Relationship," in: *The World Bank. Its First Half Century, Volume 2: Perspectives* (Brookings Institution, Washington D.C., 1997), 473–522.

Polillo, Simone, and Mauro F. Guillén, "Globalization Pressures and the State: The Worldwide Spread of Central Bank Independence," in: *American Journal of Sociology*, 110 (2005), 1764–1802.

Pomeranz, William E., "Russian Protectionism and the Strategic Sectors Law," in: *American University International Law Review*, 25/2 (2010), 213–224.

Portolano, Francesco, *Government's New Veto Powers in Strategic Sectors*, 4 April 2012, available at www.internationallawoffice.com/newsletters/detail.aspx?g=a65743ff-559b-44b5-a0e9-3fdcac3a7a36&redir=1 (status: February 22, 2013).

Prontera, Andrea, "Energy Policy: Concepts, Actors, Instruments and Recent Developments," in: *World Political Science Review*, 5/1 (2009), 1–30.

Rice University, James A. Baker III Institute for Public Policy – *Project: The Role of National Oil Companies in International Energy Markets*, available at www.bakerinstitute.org/events/the-changing-role-of-national-oil-companies-in-international-energy-markets (status: February 27, 2013).

Rios-Morales, Ruth, Dragan Gamberger, Tom Šmuc, and Francisco Azuaje, "Innovative Methods in Assessing Political Risk for Business Internationalization," in: *Research in International Business and Finance*, 23 (2009), 144–156.

Roland, Gerard, "Understanding Institutional Change: Fast-Moving and Slow-Moving Institutions," in: *Studies in Comparative International Development (SCID)*, 38 (2004), 109–131.

Rose, Peter R., "Chance of Success and Its Use in Petroleum Exploration," in: *The Business of Petroleum Exploration* (AAPG, Tulsa, OK, 1992), 71–86.

Ross, John G., "The Philosophy of Reserve Estimation," in: *SPE Hydrocarbon Economics and Evaluation Symposium Series*, 16–18 March 1997, Dallas, Texas.

—, "Risk and Uncertainty in Portfolio Characterisation," in: *Journal of Petroleum Science and Engineering*, 44 (2004), 41–53.

Ross, Michael, "The Political Economy of the Resource Curse," in: *World Politics*, 51 (1999), 297–322.

Shiravi, Abdolhossein, and Seyed Nasrollah Ebrahimi, "Exploration and Development of Iran's Oilfields through Buyback," in: *Natural Resources Forum*, 3 (2006), 199–206.

Simons, Beth A. and Lisa L. Martin, "International Organizations and Institutions," in: *Handbook of International Relations* (Sage Publications, London, 2002), 192–211.

Simpson, G. S., F. E. Lamb, J. Finch, and N. C. Dinnie, "The Application of Probabilistic and Qualitative Methods to Asset Management Decision Making," in: *SPE Asia Pacific Conference on Integrated Modelling for Asset Management*, 2000, available at www.onepetro.org/mslib/servlet/onepetropreview?id=00059455 (status: February 22, 2013).

Stevens, Paul, "National Oil Companies: Good or Bad?–A Literature Survey," in: *Centre for Energy, Petroleum and Mineral Law and Policy Working Papers*, University of Dundee, Scotland, May 2003, available at www.dundee.ac.uk/cepmlp/journal/html/Vol14/Vol14_10.pdf (status: March 1, 2013).

—, "National Oil Companies and International Oil Companies in the Middle East: Under the Shadow of Government and the Resource Nationalism Cycle," in: *The Journal of World Energy Law & Business*, 1 (2008), 5–30.

Suslick, S. B., and D. J. Schiozer, "Risk Analysis Applied to Petroleum Exploration and Production: An Overview," in: *Journal of Petroleum Science and Engineering*, 44 (2004), 1–9.

Tordo, Silvana, "Countries' Experience with the Allocation of Petroleum Exploration and Production Rights: Strategies and Design Issues," in: *World Bank Working Papers*, 2009, available at https://openknowledge.worldbank.org/handle/10986/5954 (status: February 22, 2013).

Tordo, Silvana, David Johnston, and Daniel Johnston, "Petroleum Exploration and Production Rights: Allocation Strategies and Design Issues," in: *World Bank Working Papers*, 2009, available at https://openknowledge.worldbank.org/handle/10986/5954 (status: February 22, 2013).

Tordo, Silvana, Brandon S. Tracy, and Noora Arfaa, *National Oil Companies and Value Creation* (World Bank Publications, Washington DC, 2011).

Tsui, Kevin K., "More Oil, Less Democracy?: Theory and Evidence from Crude Oil Discoveries," in: *University of Chicago Working Papers*, 2006, available at www.webmeets.com/files/papers/ERE/WC3/168/tsui-kevin-wc3.pdf (status: February 22, 2013).

Vanhanen, Tatu, *The Polyarchy Dataset*, version 1.0, available at http://www.prio.no/Data/Governance/Vanhanens-index-of-democracy/ (status: March 24, 2014).

Vedpuriswar, Ayalur V., "Managing Political Risks," in: *Global CEO. Enterprise Risk Management*, March 2002, 43–50.

Vernon, Raymond, *Sovereignty at Bay: The Multinational Spread of US Enterprises* (Basic Books, New York, 1971).

Victor, David G., David R. Hults, and Mark Thurber, *Oil and Governance: State-Owned Enterprises and the World Energy Supply* (Cambridge University Press, Cambridge, 2012).

Victor, Nadejda, "On Measuring the Performance of National Oil Companies (NOCs)," in: *PESD Stanford University Working Papers*, 64 (2007), available at http://pesd.stanford.edu/publications/nocperformance (status: February 22, 2013).

Voeten, Erik, and Michael Ross, "Unbalanced Globalization in the Oil Exporting States," in: *APSA Annual Meeting Papers*, 2011, available at http://papers.ssrn.com/sol3/papers.cfm?abstract_id=1900226 (status: February 22, 2013).

Waelde, Thomas W., "International Energy Law and Policy," in: *Encyclopedia of Energy*, 3 (2004), 557–582.

Wantchekon, Leonard, "Why Do Resource Dependent Countries Have Authoritarian Governments?", in: *Yale University Working Papers*, 1999, available at www.yale.edu/leitner/resources/docs/1999-11.pdf (status: February 22, 2013).

Warshaw, Christopher, "The political economy of expropriation and privatization in the oil sector," in: *Oil and Governance. State-Owned Enterprises and the World Energy Supply* (Cambridge University Press, New York, 2012), 35–61.

Wolf, Christian, "Does Ownership Matter? The Performance and Efficiency of State Oil vs. Private Oil (1987–2006)," in: *Journal of Energy Policy*, 37 (2009), 2642–2652.

World Bank, *A Citizen's Guide to National Oil Companies* (World Bank Publications, The World Bank Group and The Center for Energy Economics at The University of Texas at Austin, 2008).

4 A statistical analysis of oil upstream sector policies across the world

The empirical strategy of this research is a mixed-methods design, which combines statistical analysis with two case studies. The use of mixed methods serves two main purposes: first, the cross-check of findings and second, the enhancement of the understanding and the proposed explanations through the triangulation of quantitative and qualitative research. It is widely agreed that though laborious and more challenging, the application of mixed methods has a number of advantages (Axinn and Pearce 2007; Axinn *et al.* 1991; Creswell and Clark 2007; Tashakkori and Teddlie 2002 and others). In Udo Kelle's words, the mix of methods:

> can help to discover and to handle threats for validity arising from the use of qualitative or quantitative research by applying methods from the alternative methodological tradition and can thus ensure good scientific practice by enhancing the validity of methods and research findings. Or it can be used to gain a fuller picture and deeper understanding of the investigated phenomenon by relating complementary findings to each other which result from the use of methods from the different methodological traditions of qualitative and quantitative research.
>
> (Quoted in Johnson *et al.* 2007, 120)

The first part of the empirical research—that is, the statistical analysis—relies on secondary data compiled in a dataset spanning 28 countries for the timeframe 1987 to 2010. Due to limited availability of systematic data on oil production by National Oil Companies (NOCs), the econometric analysis cannot go back to the times of nationalizations in the 1970s. Yet in spite of the shorter timeframe, the trends can be identified and will be further compared with the findings of the qualitative analysis, which covers the entire time period from the 1960s/1970s until 2010. By relying on both secondary information and primary data collected through expert interviews, the case studies—as the second part of the empirical research—overcome the data limitation issues. The comparison of the statistical results with the insights provided by the case studies will help revise the proposed analytical framework of oil upstream sector policies globally.

In this chapter, the focus will lie on the statistical analysis—more specifically, the testing of the hypotheses supporting the analytical framework. First, the sampling strategy is introduced and the process of putting the dataset together is presented. Second, the discussion of data operationalization and data sources follows suit. Third, the statistical method together with its pros and cons is addressed. Fourth and finally, the results are critically discussed with respect to the proposed analytical framework.

4.1 The sample

This section first goes into the construction of the sample in reference to the conceptual and theoretical setup of this research. Second, it details the steps undertaken to manage and consolidate the data. Such a discussion is needed in order to bring transparency to the research process and ensure that the results can be replicated.

4.1.1 Sampling criteria

The unit of analysis is the oil producing country. To be defined as an oil producing country, the state has to produce crude oil domestically—however, it does not need to export it or be economically dependent on the oil sector.[1] As a starting point for the sampling strategy, the original list of Jones Luong and Weinthal's (2010) petroleum-rich countries is considered in order to map out the pool of oil producing countries worldwide. Notably, the list includes both developed and developing countries—see Appendix D in Jones Luong and Weinthal (2010, 357 ff.). To categorize a country as "petroleum-rich," the scholars looked into the relative size of its estimated petroleum resources over time as compared to other countries. Based on the Oil and Gas Journal Database, they put together three lists based on: "(1) country's position from averaging world rankings; (2) country's position from averaging quantities; and (3) country's position from weighting quantities" and finally included in the dataset those countries which were among the top 50 on two of these three lists (Jones Luong and Weinthal 2010, 310). By using Jones Luong and Weinthal's list, the implication is that petroleum-rich countries also produce their oil—in other words, that they are oil producing countries. With respect to their sampling, Jones Luong and Weinthal contended that this enabled them "to include several countries that are not usually considered petroleum-rich because they are not oil dependent, as well as to exclude countries that are usually considered petroleum-rich solely because of their dependence on oil" (2010, 310). This matches the theoretical setup of this research where full variation in oil dependence is in fact needed in order to disentangle potential drivers of the oil upstream sector policy.

Once the pool of candidates is identified, the size of the sample is determined by the data availability for the dependent variable. This is the upstream sector policy, which—within the confines of this research—is operationalized as state control ratio. In reference to Chapter 2 of this book, state control is considered to be effectively exercised through an NOC. That is why, for a petroleum-rich country from Jones Luong and Weinthal's list to be included in the sample, it first needs to have an NOC. This domestic operator is the branch through which state control is established. Admittedly, there might be petroleum-rich countries which produce domestically and have no NOC, yet to my knowledge, these cases are minor. State control in the oil upstream sector is measured as the ratio of production operated by the NOC to the total production per country by year.[2]

To pin down the petroleum-rich countries with NOCs for the sample, Petroleum Intelligence Weekly (PIW) Top 50 dataset for the timeframe 1987–2010 is used. PIW Top 50 dataset is particularly useful for this research as it provides data on the oil production volume by the NOC by year (i.e. the numerator in the state control ratio). PIW Top 50 data is a ranking of the world's 50 largest oil and gas companies, which is published yearly by Energy Intelligence. Only those NOCs are selected from the PIW Top 50, which come from a petroleum-rich country in reference to Jones Luong and Weinthal's list. The comparison of the list of "petroleum-rich" countries with the list of countries wherefrom the NOCs included in the PIW Top 50 originate, shows that the latter is a subset of the former. Notably, given that PIW is a ranking of the top 50 companies based on their oil and gas production overall, some of the NOCs are considerably smaller in terms of their oil output and larger in the gas sector—for example, Banoco in Bahrain (rank 63 in 1991), Pertamina in Indonesia (rank 66 in 2006), Kazmunaigas in Kazakhstan (rank 67 in 2001). Consequently, the sample for the quantitative analysis in this book eventually includes both small and large NOCs in terms of oil production—thereby, excluding selection bias.

Besides the fact that all the NOCs in the PIW Top 50 data are from a petroleum-rich country, these NOCs need to fulfill two additional criteria in reference to the theoretical basis of this research in order for their respective country to be included into the sample.

Criterion 1: Is state ownership over 20 percent? –> Yes/No

First, for a company to be under "state control" and thereby qualify as a "state" operator in the upstream sector with the respective oil production building the state (as opposed to private) quota, the state ownership in the company has to be more than 20 percent. The 20 percent ownership criterion legitimizes the claim that the state has decision-making power over production quotas, investment costs and pricing on the board of directors in the NOC given that non-state ownership is dispersed over many small shareholders, who cannot and do not get together to outvote the state. The 20 percent voting rights criterion has become a rule of thumb in the corporate finance literature following an article by La Porta *et al.* (1998) with over 11,000 citations. Therefore, based on the PIW rankings, all observations are selected where state ownership ranges from 21 to 100 percent (given the categories/values in the respective dataset).

Notably, the original PIW data has been checked for internal consistency across different years for the "state owned" variable and adjusted where necessary based on Wolf's (2009) data.[3] Adjustments have been made in the cases of Petrobras (Brazil), YPF (Argentina), PDO (Oman), Lukoil (Russia), Gazprom (Russia), Slavneft (Russia), and Repsol (Spain).

Criterion 2: If the case, since when has the NOC also produced overseas?

Second, for the cases where NOCs have been active not just domestically but also in the oil upstream sector overseas, the production by the NOC to the total production per country is no longer indicative of the degree of "state control" within the oil producing country. In the absence of disaggregated data on domestic versus foreign production by the NOC, all observations are dropped starting with the year when the NOC has gone overseas in oil upstream operations (if the case). It should be noted that PIW Top 50 does not detail the domestic versus overseas production by NOCs. This type of data is proprietary and corresponding data is subject to a high charge.

For this, the oil exploration and production history of each NOC in the sample was looked up. Research was primarily based on the official websites of the NOCs and if not available, on secondary sources (case studies, articles etc.). For an overview of the NOCs with oil upstream operations abroad, please see Annex 1.

All in all, the sample finally covers 28 petroleum-rich countries in the time period 1987 to 2010, all of which have an NOC and produce crude oil domestically. Since data is not available for each country for the entire period, it is an unbalanced panel. It comprises 434 observations in total. For a list of the countries catalogued in the dataset (including number of observations), please see Annexes 2 and 3.

4.1.2 Data construction

For the construction and consolidation of the dataset, there are several steps that need to be further undertaken.

Step 1: Data addition (2010) for NOC oil production

Due to the fact that the purchase of the PIW dataset was made in January 2010, the initial database on production by NOCs covers the timeframe 1987 to 2009. For more data points, the observations for 2010 were introduced manually based on the PIW Top 50, December 2011 issue.

Step 2: Data compilation

The complete dataset is compiled from several different sources, as deemed appropriate for the operationalization—for this, please see section 4.2 below.

Step 3: Correction for OPEC countries for oil production by country (including condensates)

In order to compute the state control ratio, the numerator—that is, the oil production by the NOC—is taken from PIW Top 50 data, i.e. the "output (liquids)" by Oil Company variable, and the denominator—namely, oil production by country—comes from British Petroleum's Statistical Review of World Energy 2011, i.e. the "oil production" by country variable.

The former includes "crude oil, natural gas liquids and condensates" (2009 Footnotes, PIW Top 50). While this research concentrates only on crude oil, in the absence of more appropriate data which does not include natural gas liquids and condensates, this variable is the best available fit. In turn, "oil production" from the BP dataset covers crude oil production as well as natural gas liquids (NGLs), "the liquid content of natural gas where this is recovered separately" (BP 2011, "Oil Production—barrels" Worksheet Footnotes). In order to have the same coverage both for the numerator and denominator, condensates need to be added to the denominator.

Only data for the OPEC countries has been found on condensates. However, it is considered that imputing the amount of condensates to the total production in non-OPEC countries would not really influence the ratio of state control. To compute the volume of condensates for OPEC countries, Energy Information Administration (EIA) time-series data has been used. More explicitly, given EIA's publicly available data on oil production volumes by country, both including and excluding lease condensates, the volume of lease condensates can be subtracted. This is then added to the BP "oil production" variable for the OPEC countries.

Step 4: Data consolidation

Due to the unit of analysis in this research (i.e. oil producing country), data needs to be aggregated to country level. More specifically, when there is more than one NOC operating in the oil upstream in the country, production volumes by NOC per year are weighted based on the state ownership percentage and summed up to build the total output by NOCs in a country. This way both small and large NOCs are accounted for.

Also, due to data reporting issues by NOCs, "state control" had to be adjusted to "1" for a few cases where the NOC is the sole operator in the country yet it reports more than the overall oil production per country (in which case the initial ratio amounts to values larger than 1).

4.2 Data operationalization and data sources

This section discusses the operationalization of the variables and their data sources. The dependent variable is first addressed. Given that in the existing literature there has been little previous attempt to measure upstream industry control, the advantages and limitations of the operationalization as state control ratio are presented. Second, the focus shifts to the main explanatory and control variables.

4.2.1 Dependent variable: "de facto" control as ratio

Drawing on the conceptualization of upstream sector policy discussed in section 2.2, the dependent variable seeks to capture sector participation in the oil upstream industry by the NOC (state) versus International Oil Companies (IOCs; private). The focus of this work is control—that is, operational control—where the variation actually lies. Besides Jones Luong and Weinthal's research, which puts forward a categorical variable to distinguish between forms of ownership and control, to my knowledge there is no quantitative work on the determinants of ownership and/or control structures.

To gauge state control in the oil upstream sector, as introduced before, this analysis uses the ratio of the oil production volume by the NOC to the total oil production per country (annual figures). This variable runs between 0 and 1, or in percentages, from 0 to 100 percent. The difference from 1 or respectively, 100 percent represents private control. Although Palacios did not use this measure in her work in a quantitative manner, she briefly took note of it: "another measure of the industry's openness to private investments [besides types of contracts/legal arrangements present in the upstream, *my note*] is the actual share of total production that private companies have managed to secure" (2002, 11). This corroborates the rationale behind the measure of upstream sector policy proposed in the present book—namely, the public versus private sector's share in total oil production at the oil upstream industry level.

To compute the ratio, the numerator, i.e. "oil production by NOC," is taken from the PIW dataset. So far this data has been used by only one scholar, Christian Wolf (2008, 2009), to compare economic performance of NOCs versus IOCs in the timeframe 1987 to 2006. The denominator for the ratio comes from British Petroleum's Statistical Review, which is open source. Due to its differences in coverage as compared to the numerator, condensates had to be added from the EIA database, "World Crude Oil Production" (1970–2009), as discussed in section 4.1.2.

Importantly, this ratio of state control captures operational or "de facto" control, as compared to "de jure" control in Jones Luong and Weinthal's work (2010). The "ownership structure" variable in Jones Luong's self-compiled dataset is coded based on the country's legal framework, which may include the constitution/s, official laws, regulations governing the energy sector and where available, legal contracts between state and private entities in the petroleum sector (Jones Luong and Weinthal 2010, 311). This variable thus covers what is permitted based on the legislation in place—i.e. "de jure" ownership and control. Furthermore, it is categorical and may take one of the following four values:

1: S1 = state ownership with state control;
2: S2 = state ownership with foreign control;
3: S3 = domestic private ownership with control;
4: S4 = foreign ownership.

Fifty petroleum-rich countries in the developing world in the timeframe 1900 to 2005 are coded in Jones Luong's dataset. Despite the larger sample of the "de jure" control variable, there are a number of limitations to it. The comparison of the data for "de jure" control with that for "de facto" control—when using the same setup of 28 countries for 1987 to 2010 (see section 4.1 above)—shows that variation in "de jure" control is very limited, in particular within countries but also across countries. By contrast to "de facto" control where even small changes can be captured by the ratio, the "de jure" control variable builds slightly more than 28 data points corresponding to the number of countries in the sample, few of which changed legislation and more importantly, permissibility of foreign involvement in the oil upstream over the course of the analyzed 24 years. This makes it impossible to run a time-series cross-sections regression. Additionally, when comparing the data on "de jure" with that on "de facto" control for the 28 countries, the codings for 10 countries do not fit the empirical reality—namely, in the cases of Algeria, Argentina, Colombia, Ecuador, Iran, Libya, Nigeria, Qatar, Syria and Venezuela. In other words, there is a mismatch between what is potentially permitted ("in theory") and what the reality looks like ("in practice"). On these grounds, this book—in its aim of coming close to reality—acknowledges the relevance of "de jure" control for the documentation of the rules of the games, but prefers the operationalization of upstream sector policy as "de facto" control.

4.2.2 Independent variables

The five independent variables will be introduced with respect to the explanations proposed in the analytical framework and the underlying hypotheses.

To start with the *geological conditions*, it is worth noting that data on oilfield characteristics, including upstream cost data, is proprietary to oil producing countries and energy consulting companies. Publicly available data such as oil "proven reserves" or "estimated reserves" only covers size and not the degree of complexity involved in the extraction process. Tsui (2010) argues along the same lines with respect to mineral endowment.

Previous studies propose field location, i.e. whether onshore or offshore, as an indicator for the difficulty of upstream operations (Jojarth 2008; Ross 2006). The development of offshore oilfields is associated with higher costs (Hamilton 1983; Kaiser 2007; Le Leuch and Masseron 1973) even though technological advancements have significantly shrunk the gap between onshore and offshore in terms of cost differences in the past three decades (Babusiaux 2004).

To test Hypothesis 1 on the impact of geological conditions, two measures are used. First, an "onshore/offshore" variable is going to be factored into the regression analysis in line with the idea that offshore operations involve a higher level of complexity. As compared to both Jojarth (2008) and Ross (2006), who include a dummy variable, this chapter calculates a ratio of offshore fields out of the total number of oilfields in the country, based on *Oil and Gas Journal*'s "Historical Worldwide Oil Field Production Survey." This dataset contains a dummy variable called "offshore," which documents the types of all oilfields per country in the time period 1980 to 2007. Even though the time range for this variable stops in 2007, it can be assumed that the new discoveries between 2008 and 2010 did not change the ratio of offshore to onshore oilfields considerably (if at all).

The second measure of geological conditions—to be used in this analysis—is the average depth of oilfields in a country. This is based on the "depth" variable from the same source, Oil and Gas Journal Database. Tsui (2010) relies on the same variable to account for cost-determining oilfield characteristics in his study on the impact of oil wealth on democracy. Though rough, this measure is yet another estimate of geological conditions given the common wisdom that higher depths assume more complexity within upstream operations.

When it comes to the *technical capabilities of the NOC*, there is no publicly available data on capabilities of NOCs nor level of technical/technological know-how.[4] Therefore, proxies need to be used. Arguably, the NOC is more likely to dispose of technical capabilities in countries with higher state capacity. Drawing on Tilly, state capacity stands for "the extent to which the governmental agents control resources, activities, and populations within the government's territory" (2003, 41). It is assumed that the countries which can control and capitalize on their resources are also the ones more likely to develop the technical capital of state-owned enterprises. Nevertheless, due to the specificities of oil producing countries, their rentier status and ruling bargain, only some of the indicators proposed by this scholarship can be applied.

One of these proxies is GDP per capita. The rationale behind the use of this proxy is that a wealthy state is likely to be technically more capable since it can afford not only acquiring cutting-edge technology but also training its people from NOCs. Fearon and Laitin consider GDP per capita "a proxy for a state's overall financial, administrative, police, and military capabilities" (2003, 80). Although this measure has some serious limitations because it includes oil rents, may cover more than just technical capabilities, and the results might be thus biased, its inclusion is arguably worthwhile.

A second proxy to capture NOC technical capabilities is an education measure, namely "school enrollment, tertiary." Tertiary education, whether or not to an advanced research qualification, follows upon completion of education at the secondary level and can be considered a fair indicator for the capabilities of a firm. Since the NOC is the major employment provider in petroleum-rich countries, the educational attainment is likely to be reflected by the employees of the state operator. The same approach to educational attainment has been used by other researchers like Blalock and Gertler (2008) and Burki and Terrell (1998). To compute "school enrolment, tertiary" in numbers (thousands), "school enrollment, tertiary (percent gross)" from the World Bank's Education Statistics Database and "population, total" from the World Bank's World Development Indicators have been multiplied.

For the *international oil price* variable, the level of oil prices (in U.S. dollars, adjusted to 2010) is taken from BP's Statistical Review of World Energy 2011.

With regard to *oil reliance* or synonymously, oil dependence, scholars have used different measures. The World Bank study (Tordo *et al.* 2011, 46) discusses how "national dependency on the petroleum sector" affects NOC value creation. Drawing on the theoretical discussion of rentierism and the role this might play for the policies shaped and adopted by the ruling elites, the variable used in this analysis needs to account for the extent to which the bulk of state revenues comes from the petroleum sector or instead, from alternative sources. Oil wealth—operationalized as "oil rents (percent of GDP)"—is in this respect a good proxy of national oil reliance. The indicator comes from the World Bank Development Indicators and is defined as "the difference between the value of crude oil production at world prices and total costs of production." The same variable of oil wealth is used by other scholars to account for the level of dependence on petroleum rents (Bastiaens 2011; Farzanegan 2012; van den Bosch 2012).

As a second proxy, oil export dependence—oil exports as a percentage of GDP—may also be a good fit because it taps both into the relevance of petroleum as a source of export revenue and its relative importance for the national economy (Ross 2001, 338). To calculate this, data both for the numerator and the denominator is taken from the IMF World Economic Outlook 2011, i.e. TXGO: "value of oil exports" (billions U.S. dollars) and NGDPD: "gross domestic product, current prices" (billions U.S. dollars). A similar approach to oil reliance has been adopted by Leite and Weidmann (1999), Ross (2001), Sachs and Warner (1995).[5]

Finally, to test for the effect of *executive constraints*, the political constraints index (POLCONIII) from Polcon Dataset (2010) is considered. This is a composite measure of the extent to which the government is constrained in its actions when willing to change policy (Henisz 2005, 3). The index has been widely used to proxy for political risk (e.g. Andonova and Diaz-Serrano 2009; Click and Weiner 2010) and policy stability (Vaaler and Schrage 2009). While the index is well in line with the theoretical argument supporting Hypothesis 5, as it measures the feasibility of change in government policy with regard to the number of veto points and preference heterogeneity, the downside is that preferences are coded in terms of party preferences (party composition of legislatures). This turns it into a relatively rough indicator, incapable of capturing fine variation in a region like, for example, the Middle East, where parties have a lesser role to play and party composition may be homogenous, but where institutional constraints on government authority do vary considerably across countries.[6]

Due to the inherent assumption in the analytical framework regarding the primary goal of state leaders and their potential engagement in rent-seeking activities to secure this, *corruption* cannot be left out of the statistical analysis. For the sake of political survival, patronage under different forms needs to be dispensed. This encourages corruption and might make state leaders more prone to support the option of an NOC-dominated upstream because this way private rent-seeking is more facile to pursue and camouflage.[7] It is, however, not clear if high corruption leads to more state control in the upstream or if it is the state control in the upstream which paves the way to more rent-seeking opportunities and thus, higher corruption. Arguably, the causation can work either way. Moreover, under provision that executive constraints are low, state leaders might prefer to support their patronage network through funds raised from IOCs, which are known to be more efficient and could thus generate higher profits. Supposing that corruption leads to state control in the upstream and not the other way round, it is not clear whether it brings about more or less state control.

Consequently, given the theoretical assumptions in the analytical framework, the corruption variable will be exploratorily included in the analysis. Data on "control of corruption" is taken from Political Risk Services (PRS)—International Country Risk Guide (ICRG) reports. Drawing on Jeong and Weiner (2012), this is a measure of the demand side of corruption, which is meant to characterize the institutional environment of the host country. The PRS–ICRG corruption variable is a good match for the theoretical set-up of this research as it maps out "actual or potential corruption in the form of excessive patronage, nepotism, job reservations, 'favor-for-favors', secret party funding and suspiciously close ties between politics and business" (PRS–ICRG T4B—Financial Risk). This can distort "the economic and financial environment; it reduces the efficiency of government and business by enabling people to assume positions of power through patronage rather than ability; and, last but not least, introduces an inherent stability into the political process" (ibid.). The same PRS–ICRG data has been used by many scholars (e.g. Busse and Hefeker 2007;

Gupta *et al.* 2001; Tavares 2003) given that it spans a long time period by comparison with other corruption indicators (World Bank 2009, 26).

4.2.3 Control variables

To take note of potential alternative explanations, control variables are factored into the analysis. These are: international institutions, regional effect, policy inertia, and ownership structure of the NOC. The former three are also considered in Jones Luong and Weinthal (2010)'s statistical chapter.

In reference to the discussion about the role of *international institutions*, the effect of OPEC needs to be accounted for. More specifically, a self-coded dummy variable for OPEC membership is included in line with the argument that in the 1970s OPEC encouraged developing countries to nationalize their oil resources and might have retained some influence on the upstream sector policies of its member states thereafter. Yet it is worth mentioning that both scholars and practitioners have emphasized the declining role of OPEC across oil producing countries during the past decades (Al-Chalabi 2003; Griffin and Xiong 1997).

For the *regional effect*, a self-coded dummy variable for Middle East and North African (MENA) countries is used. This should not only capture potential contagion across MENA countries in terms of their upstream policies but also account for the regional specificity.

The *policy inertia* or path dependence appears inherent in the effort to explain policies. Güven's definition of institutional stickiness or resilience is edifying as to the reasons underpinning the phenomenon:

> Resilience can be defined as a structural disincentive to change a given arrangement on the part of actors capable of inducing such change. It is typically attributed to the endogenous features of an institutional domain. All institutions are resilient to some extent, for institution creation involves sunk costs and is followed by a degree of adaptation by actors, rendering change unattractive in principle.
>
> (2009, 167)

While institutions are different from policies (as introduced in Chapter 2), they share the element of stickiness. Similarly to institutional changes, policy changes are associated with political and economic costs for certain actors or powerful interest groups in the respective policy domain (Thelen 2003). The widespread position of historical institutionalists in this regard is that policy arrangements first acquire their relevance through the instrumental function they fulfill and, in time, grow in stickiness through the self-reinforcing mechanisms they generate, such as increase in returns (despite possible inefficiency), in power or legitimation for a select group of actors (Mahoney 2000; Pierson 2000). As a measure of policy inertia or path dependence, many scholars have used a lag of the dependent variable—for example, Janssen and de Vries (2000) in their study of climate change policy, Lewis (2009) or Lledó and Poplawski-Ribeiro (2011) in their approach to fiscal policy.

Finally, given the variety of NOCs, there needs to be disentangled between NOCs where the state has 100 percent of the equity shares (own 1), where the state is the majority owner, i.e. over 50 percent (own 2) or where the state only has a minority of the shares in the company yet over 20 percent (own 3) in line with the first sampling criterion spelled out in section 4.1.1. Scholars have shown that depending on the firm ownership structure, NOCs can behave quite differently (Jeong and Weiner 2012). Consequently, it has to be taken note of the *ownership structure of the NOCs*. "Own 1," "own 2" and "own 3" are all dummy variables self-coded based on the PIW dataset (1987–2010). "Own 2" (majority state-owned NOC) is the base.

For the data in a nutshell: please see Annex 4 for an overview of the variables including both data operationalization and data sources, Annex 5 for the univariate statistics and Annex 6 for the correlation matrix.

4.3 Statistical tools

The limited range of the dependent variable suggests a censored regression model (Maddala 1983; Wooldridge 2001, 2009). Annexes 7–9 offer a graphical representation of the data. Several characteristics of the data— namely, (1) the fractional nature of the dependent variable as in state control ratio; (2) the inclusion of a lagged dependent variable due to policy inertia or path dependence (dynamic model); (3) the unobserved heterogeneity across oil producing countries; and not least, (4) the unbalanced panel for the 28 countries in the timeframe 1987 to 2010—recommend the use of a two-limit or doubly-censored Tobit model with the dynamic panel fractional (DPF) estimator (Elsas and Florysiak 2010, 2012).

Alternatively, applying the fixed effects (FE) approach with a dynamic panel Tobit model would disregard the fractional nature of the dependent variable and, based on Chang (2004), render the estimates inconsistent. Loudermilk shows that the difference between fractions (and percentages) and level values[8] from an econometric stance is that the former "are not probabilistic outcomes, yet they have both two-corner solution outcomes and continuous outcomes in the interval (0, 1). Consequently, most standard models are inappropriate for estimation" (2007, 462). Loudermilk's estimation of firm dividend policy proves that overlooking the doubly-censored nature of fractional responses and the dynamics in the model brings about misleading results (ibid., 467ff.).

Based on an example from capital structure research, Elsas and Florysiak (2010, 2012) also prove that alternative estimators in a dynamic and unbalanced panel data setup with a fractional dependent variable are severely biased. More specifically, the FE estimator is biased because the introduction of a lag of the dependent variable leads to correlation between the error term and the independent variables, which makes the FE biased for fixed T. The same applies to the pooled OLS. Both of these observations were also confirmed by Loudermilk (2007, 463). As for the GMM estimator proposed by Blundell and Bond as well as other instrumental-variables fixed effects estimators, these may be consistent with a lagged dependent variable in the panel setting but "are potentially biased since they do not take the fractional nature of the dependent variable into account" while "the impact of a fractional dependent variable on the estimators' properties is largely unknown" (Elsas and Florysiak 2012, 8). In short, the DPF estimator is designed to be both unbiased and consistent in the context of an unbalanced and dynamic panel with a ratio as the dependent variable and unobserved heterogeneity. Based on Monte Carlo simulations, John *et al.* (2012) endorse Elsas and Florysiak's findings as to the DPF estimator.

The need to develop special models for fractional dependent variables has been recognized by several scholars. In fact, the DPF estimator proposed by Elsas and Florysiak builds on the work of Papke and Wooldridge (2008) and Loudermilk (2007). While the former approach allows for endogenous explanatory variables and unobserved heterogeneity, the model cannot be applied when including a lagged dependent variable. In turn, the latter proposes a doubly-censored Tobit with a lagged dependent variable and unobserved heterogeneity, but only for a balanced panel. Starting from Louldermilk's (2007) model, the DPF estimator has a changed specification of the fixed effects distribution in order to allow both for unbalanced panel data and a lagged dependent variable (Elsas and Florysiak 2012, 9). The DPF estimator has been used just in corporate finance research so far (13 citations). Yet, its specification and rationale are well applicable in other fields, as its developers also suggest.

Given the small change in state control within countries across years, the DPF estimator is a good match since it can capture subtle changes of adjustment in the dependent variable. Based on the doubly-censored Tobit specification, it also allows for corner solution observations at both zero and one. While the sample has no observation where state control is 0, in 48 out of the 434 observations state control equals 1, i.e. in approximately 11 percent of the observations.

Once the advantages of using the DPF estimator in the context of the state control ratio in an unbalanced and dynamic panel have been spelled out, the one potential downside of the DPF estimator needs to be emphasized as well. In case of misspecification of the model through omission of an explanatory variable or inclusion of a superfluous regressor, the estimates of the DPF model may be slightly biased because of the presumed parametric specification of the fixed effects distribution. However, the simulation tests show that the bias is only minor and the DPF estimator remains robust (Elsas and Florysiak 2012, 23).

Along the lines of potential misspecification, the inclusion of a lagged dependent variable has its pros and cons. Achen (2000) shows that the inclusion of an autoregressive term as a control in a non-stationary setup might not only render it statistically significant and improve the fit of the model, but at the same time diminish the value of the substantive variables and even render their coefficients statistically insignificant. While the scholar points to the fact that "obtaining truly adequate coefficient estimates and standard errors in panel studies or any other time series context with autoregressive terms and serial correlation is no simple matter" (ibid., 25), he offers no advice on what to use in the presence of serial correlation. The lagged dependent variable might lower or suppress the explanatory power of other independent variables, yet for the testing of dynamic theories "any model that omits such a dynamic component is under-specified" (Keele and Kelly 2006, 190). A truism among statisticians, an omitted variable leads to biased estimates. In a Monte Carlo analysis, Keele and Kelly show that the downward bias of the estimates pertaining to other independent variables in the presence of a lagged dependent variable is negligible when the sample size is sufficiently large. More explicitly, "even when the sample size is 25 cases the bias is approximately 6% and falls to around three percent once N increases to 75" (ibid., 203). Given the size of the sample at hand, it turns out that the bias of the substantive coefficients is of limited concern. Instead, the omission of the lag—given the addressed relevance of policy inertia for the economic policy adopted in the upstream sector—would be more of a serious problem.

In conclusion, a doubly-censored Tobit analysis with the DPF estimator will be applied. For the implementation in Stata 12 based on Elsas and Florysiak's recommendations (2010, 40), the *xttobit* command will be used with the following options: a Gauss-Hermite quadrature (*intm(gh)*), a lower limit at zero (*ll(0)*) and an upper limit at one (*ul(1)*). The maximum likelihood estimation of the Tobit model with DPF is based on the following latent variable specification:

$$y_{it}^{\#} = Z_{it}\varphi + \rho y_{i,t-1} + c_i + u_{it}$$

with Z_{it} a set of exogenous regressors, $\rho y_{i,t-1}$ lagged dependent variable with its coefficient, c_i the unobserved heterogeneity (i.e. the fixed effects), and $u_{it} \sim N(0, \rho_u^2)$ the error term.

The observable doubly-censored dependent variable with two possible corner outcomes is:

$$y_{it} = \begin{cases} 0 & \text{if} \quad y_{it}^{\#} \leq 0 \\ y_{it}^{\#} & \text{if} \quad 0 < y_{it}^{\#} < 1 \\ 1 & \text{if} \quad y_{it}^{\#} \geq 1. \end{cases}$$

(Elsas and Florysiak 2010, 10f.)

4.4 Results

In 25 different models, annexes 10–13 report the estimates of the Tobit analysis with the DPF estimator. The testing of hypotheses supporting the analytical framework proceeds stepwise. The geological–technological explanations (H1 and H2) are tested in models 1 through 9 both individually and combined. The oil price explanation (H3) together with the oil reliance (H4), both of economic nature, follow suit in models 10 and 11. The politico-institutional argument (H5) is tested in models 12 through 17. Finally, all the factors are factored in together and tested simultaneously in models 18 through 25. The computation of marginal effects is based on the probability that the state control ratio will range between 0 and 1 by using the *pr (a, b)* option of predict (Baum 2006, 262ff.).

4.4.1 The context

The coefficient estimates on both ratio offshore and average depth—both of them proxies for *geological conditions*—are consistently negative and robustly significant, with associated p-values of less than 0.01. The only instance when one of the geological proxies loses its significance is when introducing the technical capacity proxy of school enrolment. Yet even in these models (6 and 7), ratio offshore remains highly significant and preserves the expected sign. The models plugging all the arguments together (models 18–25) further confirm the role of geology in shaping the upstream sector policy. While ratio offshore is strongly significant (99 percent confidence interval) across all models, the other proxy for geological complexity, average depth, proves to be statistically significant in the absence of ratio offshore in the regression (despite the low correlation between the two, i.e. 0.1138).

All in all, *Hypothesis 1* receives strong support across models and thus confirms that more complex geology is likely to make oil-rich countries loosen up the state control in the upstream. This finding corroborates the quantitative results of Nolan and Thurber, who use water depth and previous finds to test for the effect of the "frontier" character of geology on the choice of NOCs vs. IOCs, and uncover that "the strong preference against NOCs [with the exception of Petrobras and Statoil, *my note*] in deeper water is roughly consistent across all four time periods" (Nolan and Thurber 2012, 159)—that is, for the entire time period under analysis, 1970 to 2008.

Based on the computation of the adjustment factor for the Tobit estimates in the present setup, it turns out that conditional on the state control ratio being between 0 and 1, 1,000 m in depth—starting from the mean values of all variables—is estimated to decrease expected state control by 1 to 2 percent. Given that the mean of average depth is about 7,500 m and the standard deviation approximately 1,800 m, the impact of the oilfields depth is not of large magnitude, yet clearly present. As for the ratio offshore, the partial effect at the mean values of the covariates is also small; namely, an increase by 50 percent (0.5 in ratio offshore) is estimated to bring about a decrease of approximately 5 to 6 percent in state control. Complex geology thus influences the oil upstream sector policy.

In turn, the relation between the *technical capabilities of the NOC* and state control is quite ambiguous. When testing the geological–technological explanations individually, the only model where the coefficient estimate is positive and strongly significant is model 5, when GDP per capita is singularly tested in the presence of control variables. In the models 3, 4, 7 through 9, the negative sign on both the educational and wealth proxy of technical capacity (school enrolment and GDP per capita, respectively) runs counter to the expectation. Yet when factoring in the arguments together, though low-magnitude and statistically significant at a p-value of 0.1, the coefficient estimate on the educational proxy is positive (model 18), as expected in Hypothesis 2. Nonetheless, the coefficient on GDP per capita remains either statistically insignificant (model 24) or stays negative at 1 percent significance level (model 22).

The change of sign for school enrolment is likely to be caused by the introduction of the interaction term between political constraints and oil rents of GDP. The negative strong correlation between school enrolment and oil reliance can flip the sign on the school enrolment coefficient when simultaneously testing the arguments.[9] The comparison of the log-likelihood across models points out that this decreases dramatically when the school enrolment proxy is used.

Admittedly, the school enrolment proxy might not be the best fit for the concept of NOC technical capabilities. Given that in many oil producing countries, particularly in the MENA region,[10] the school system, let alone the university system, has been developed only recently, the educational attainment at the tertiary level might not be reflective of the technical capital at the disposal of the NOC. With prevalence in the geological and more generally, technical ambit, many individuals from the MENA producer countries (at least) have pursued engineering degrees abroad. Consequently, the enrolment in domestic universities might be eventually a poor indicator of an NOC's technical capacities. These can be raised by employing nationals with degrees from abroad, supporting their well-targeted education overseas and bringing them back into the company afterwards, or by acquiring foreign expertise. The same may be the case with other oil producing regions.

In turn, for the GDP per capita, it is likely that the coverage of this proxy goes beyond the technical capacities of the NOC. Models 8 and 9 present a strongly significant yet negative relation between GDP per capita and state control. While model 22 upholds the negative sign and level of significance, when including the MENA variable in model 24 to test for the regional specificity, GDP per capita variable is no longer statistically significant. When significant, the findings largely show that the higher the GDP per capita is, the lower the state control in the oil upstream sector becomes. This can testify for the research on economic performance of NOCs vs. IOCs (e.g., Al-Mazeedi 1992; Al-Obaidan and Scully 1992; Eller *et al.* 2011; Victor 2007). More specifically, more wealth—as in higher GDP per capita— is associated with an upstream sector run by private operators, which are likely to be more efficient and generate more revenues for the host countries. Additionally, higher wealth is oftentimes associated with more liberalized economies, which would also explain the finding.

All in all, in the absence of better indicators for the technical capabilities of the NOCs and given the available data, *Hypothesis 2* does not raise enough support to be confirmed.

Regarding the *oil price* hypothesis, contrary to the prediction, the findings show that the oil price level negatively affects the state control ratio at very high confidence levels. The relationship holds across all model specifications.[11] The plot in Annex 9, "State control ratio (mean) versus time," clarifies this. The time period of highest state control appears to be the very late 1990s, when prices were low. What is also clear is that higher oil prices in the past few years have not been associated with higher state control.

Despite the strong positive relation between oil prices and expropriations reported by several empirical studies (e.g., Guriev *et al.* 2009, 2011; Jones Luong and Weinthal 2010; Stroebel and van Benthem 2010; Warshaw 2012), the present statistical analysis shows that when leaving the period of nationalizations (1970–1980) away, the coefficient on oil price is no longer positive as expected. Even when using the log of the oil price,[12] the coefficient on the log remains negative and statistically significant at 1 percent confidence level. This shows that beyond the 1980s, state control does not increase with higher oil prices. The sign on the coefficient is in fact negative. A speculative explanation for this relates to depletion policy. More explicitly, in spite of high oil prices, oil producing states expect even higher rises and do not wish to take their oil out of the ground until at a later point in time for the sake of even larger windfalls. *Hypothesis 3* is thus disconfirmed.

In conclusion to the contextual factors, while the relevance of geological conditions has been proven, the other two hypotheses concerning the NOC's technical capabilities and the oil price cannot be confirmed—the latter even presented the opposite effect than expected.

4.4.2 Domestic constraints

In the following, the focus moves to the two main explanatory variables— namely, oil reliance and executive constraints. Their significance and effect will be tested both individually and in interaction.

To start with, Model 10 elicits some evidence for the *national oil reliance* argument in Hypothesis 4. While the coefficient on oil rents of GDP is statistically significant at the 10 percent level and of positive sign, the oil exports variable in model 11 remains insignificant. Furthermore, in model 18, if interpreted singularly, oil rents of GDP have a significant positive effect on state control at 10 percent significance level when there are no political constraints, or in other words, when the political constraints index equals zero. Substantially, this relation is quite meaningful at a visual inspection of the plot of political constraints against state control. The graph (Annex 14) shows a considerable number of zeros on political constraints.

When computing the marginal effect of oil rents of GDP,[13] it turns out that an increase by 50 percent in oil rents may bring about an increase of 5 percent in state control at the average values of the covariates. Given the historical changes in oil rents, this is not a negligible finding. As for the effect of oil exports of GDP as a second proxy for oil reliance, the computation of marginal effects renders the coefficient statistically insignificant.

Given the empirical evidence, *Hypothesis 4* thus raises partial support.

The most noticeable aspect of the results overall is the positive and robustly significant coefficient estimates on *executive constraints*, with associated p-values of less than 0.01 across all models. Though counterintuitive at first, higher political constraints in oil-rich countries are conducive to higher state control in the upstream sector. If an oil-rich country went from no checks and balances to extensive checks and balances—from one extreme to the other—the expected increase in state control at the mean values of the covariates would be between 10 and 17 percent. Even though such shifts are rare, the effect of political constraints on state control ratio is significantly present and positive, as expected in Hypothesis 5.

Studies like Guriev *et al.* (2009, 2011), Warshaw (2012) or Christensen (2011) looking into the determinants of expropriations have uncovered the contrary, i.e. that more checks and balances are associated with fewer expropriations. Yet this is not contradictory because, as explained before, nationalizations/expropriations and increase in state control are related but not similar phenomena.

Based on the present analysis, more constraints on state leaders in oil-rich countries are likely to result in expressions of resource nationalism meaning more state control in the upstream sector, which thus confirms *Hypothesis 5*.

Notably, the interaction term between political constraints and oil rents of GDP—accounting for the interaction between executive constraints and oil reliance—is statistically significant at a p-value of less than 0.01 across all the model specifications. A robustness check with oil exports of GDP accounting for oil reliance renders it statistically significant in model 25 yet not in model 19. This might have to do with the fact that in model 19, oil exports of GDP are tested together with the school enrolment variable and the correlation between the two is quite high (–0.57). Yet all the other models (14–18 and 20–25) show that the interaction term is strongly significant and furthermore, that oil reliance has a moderating effect when the level of executive constraints is low—more specifically, low level of executive constraints leads to more state control when oil reliance is high and to less state control when oil reliance is low.[14] These results substantiate the fact that domestic institutions have an important role to play in explaining upstream sector policies and that their effect is furthermore shaped by the degree of oil reliance in the respective countries.[15]

As concerns corruption, the PRS variable does not reach the level of statistical significance in any model specification.

4.4.3 Alternative explanations

The inclusion of the control variables renders the alternative explanations relevant. More specifically, the lag of the dependent variable is strongly significant across all model specifications. This shows that there is a clear element of path dependency or policy inertia, which supports previous statistical findings as well as anecdotal evidence provided by Ahrend and Tompson (2007), Bellin (2004), Marcel (2006) and others.

The ownership structure of the NOC is also relevant to the control structures of the oil upstream sector. In particular, it is worth noting that "own 3" (minority state-owned NOC) has a reductive effect on the state control ratio (as compared to the base: majority state-owned NOC) and is significant at a p-value of less than 0.01 throughout most model specifications.

Furthermore, the coefficient on OPEC has a positive sign and is largely significant (most of the time, at 1 percent significance level), which confirms that countries which are members in this international organization are expected to have higher state control in the oil upstream sector.

Finally, a dummy for the MENA region was added to some of the models to account for the potential exceptional character of the oil producing countries from the region. The models which include the dummy variable render it consistently significant at 1 percent significance level. The computation of marginal effects shows that MENA countries are indicative of 7 to 14 percent more state control than the rest of the oil-rich countries. This testifies for the propensity of the region towards resource nationalism and larger state role in the economy, which corroborates previous studies like Crystal (1995), Herb (2009) and Tétreault (2008).

4.5 Conclusions to the statistical analysis

The statistical testing of the analytical framework proves that the context shaped by geology, technical capabilities of the NOC and international oil price matters only partly for the oil upstream sector policy chosen by the state leader and/or the collective executive. In turn, the degree of rentierism (or synonymously, the extent of oil reliance) and even more so, the efficiency of domestic political constraints define the state leader and collective executive's room for maneuver in their policy-making for the oil upstream sector.

Complex geology has only a small constraining effect on the extent of control pursued by the state in the upstream sector. Difficulty of extraction posed by offshore fields and operations at higher depths reduce the executive's appetite for state control over the upstream sector only marginally. By comparison, the technical endowment of the national operator does not prove relevant to the upstream control structures at all. However, this might be a consequence of the operationalization employed for the NOC's technical capacity, for which data is proprietary to NOCs and proxies are (arguably) only partial measures of the actual concept. As for the effect of international oil markets, upstream sectoral control turns out to be insensitive to oil prices. Contrary to the existing scholarship on nationalizations, which are yet different from increase in state control, the impact of oil prices seems to be negative. In short, apart from the geological characteristics of the oil-rich country, neither of the other contextual factors turns out to make state leaders more lenient so as to grant IOCs production rights.

In turn, the main arguments put forward by the analytical framework prove to have quite high explanatory power. More specifically, a high degree of oil reliance is estimated to increase the state leader's proclivity for more state control in the oil upstream sector even though the impact is still of low magnitude. The most significant results are obtained for the extent of executive constraints and the interaction between this and the country's level of oil reliance. Strong support has been found for the high influence of efficient political institutions in keeping the "national patrimony"—as in oil resources—under state control. Given the ambiguity of the existing studies about the effect of political institutions, this work shows that in the context of oil-rich countries, more limitations on the executive results in more state control in the upstream. Moreover, the interaction term, which is statistically significant at very low p-values, accounts for the fact that in the absence of executive constraints and the presence of high oil reliance, state leaders opt for increased production through the NOC to raise more funds for the sake of political survival. Yet when the degree of national oil reliance is low and executive constraints non-existent, they might as well focus on other economic sectors as sources of income and give in rights to (foreign) private firms in the oil upstream.

All in all, the statistical analysis largely validates the proposed analytical framework. In-depth insights are further gathered within the case studies, the quantitative and qualitative methods will be thus triangulated and the results cross-checked before any final conclusions are drawn.

Notes

1 A country is considered oil-dependent if at least 40 percent of the GDP comes from oil revenues (Luciani 1990; Ross 2012).
2 For a justification of this measure, see section 4.2.1 below.
3 Thank you to Christian Wolf for sharing his data (2009) with me.

4 "This is all proprietary data," to quote a Saudi Aramco representative in an interview on December 13, 2011.
5 Notably, the way countries are selected in the sample is *not* based on oil dependence. Instead, they are petroleum-rich countries—see definition in section 4.1.1.
6 The executive constraints index ("xconst") from Polity IV Dataset (1800–2009) might have been another potential candidate for the politico-institutional explanation. It is however highly correlated with the polcon index (approximately 70 percent), so it does not really matter which one is used. Additionally, the executive constraints index is more highly correlated with several other explanatory variables than the polcon index—hence, the preference for the latter.
7 This is favored by the lack of transparency in unlisted companies, which do not need to disclose their finances using international accounting standards and further regulation. According to Transparency International (2011), NOCs score low on transparency.
8 A good example here is market share versus market size (Loudermilk 2007).
9 The correlation threshold is taken at 0.6, which is admittedly quite high and thus the issue of multicollinearity can be raised in the case of school enrolment—oil reliance (–0.57/–0.51).
10 12 out of the 28 countries in the sample come from the MENA region.
11 The only exception is model 22 when the oil price—statistically insignificant this time—is tested together with GDP per capita. Despite low correlation, it may be the case that the effect is absorbed by the GDP per capita coefficient.
12 Results are available upon request.
13 The *margins . . ., predict (e (0, 1))* command in Stata 12 (Stata Reference Manual online).
14 Both sign and significance are confirmed when using *margins* to test for the marginal effects of the interaction term.
15 Plots of the interaction are available upon request.

References

Achen, Christopher H., "Why Lagged Dependent Variables Can Suppress the Explanatory Power of Other Independent Variables," in: *Working Paper Princeton University*, 2000, available at www.princeton.edu/csdp/events/Achen121201/achen.pdf (status: February 21, 2013).
Ahrend, R., and W. Tompson, "Realising the Oil Supply Potential of the CIS: The Impact of Institutions and Policies," in: *Econ Papers*, 484 (2007).
Al-Chalabi, Fadhil J., *US Encyclopedia of Energy – A History of OPEC* (Centre for Global Energy Studies, London, 2003).
Al-Mazeedi, Wael, "Privatizing National Oil Companies in the Gulf," in: *Journal of Energy Policy*, 20 (1992), 983–994.
Al-Obaidan, Abdullah M., and Gerald W. Scully, "Efficiency Differences between Private and State-owned Enterprises in the International Petroleum Industry," in: *Applied Economics*, 24 (1992), 237–246.
Andonova, Veneta, and Luis Diaz-Serrano, "Political Institutions and Telecommunications," in: *Journal of Development Economics*, 89 (2009), 77–83.
Axinn, William, Thomas E. Fricke, and Arland Thornton. "The Microdemographic Community-Study Approach Improving Survey Data by Integrating the Ethnographic Method," in: *Sociological Methods & Research*, 20/2 (1991), 187–217.

Axinn, William, and Lisa D. Pearce, "Motivations for Mixed Method Social Research," in: *Mixed Method Data Collection Strategies* (Cambridge University Press, Cambridge, 2007), 1–10.

Babusiaux, Denis, *Oil and Gas Exploration and Production: Reserves, Costs, Contracts* (Editions Technip, Paris, 2004).

Bellin, Eva, "The Robustness of Authoritarianism in the Middle East: Exceptionalism in Comparative Perspective," in: *Comparative Politics*, 36/2 (2004), 139–157.

Bastiaens, Ida, "Variations in Foreign Direct Investment in Authoritarian Regimes," in: *2011 APSA Annual Meeting Papers*, available at http://papers.ssrn.com/sol3/papers.cfm?abstract_id=1900183 (status: February 21, 2013).

Baum, Christopher F., *An Introduction to Modern Econometrics Using Stata* (Stata Corp, College Station, TX, 2006).

Blalock, Garrick, and Paul J. Gertler, "Welfare Gains from Foreign Direct Investment Through Technology Transfer to Local Suppliers," in: *Journal of International Economics*, 74 (2008), 402–421.

van den Bosch, Marie Aliénor, "Natural Resources, Economic Performance and the Politics of Public Finance in Rentier States," in: *Princeton University Working Papers*, 2012, available at www.princeton.edu/politics/graduate/courses/seminars/comparative-politics-semi/MAvdB_POL591Paper_MAY4.pdf (status: February 21, 2013).

British Petroleum, *BP Statistical Review of World Energy*, June 2011, available at www.bp.com/assets/bp_internet/globalbp/globalbp_uk_english/reports_and_pub-lications/statistical_energy_review_2011/STAGING/local_assets/pdf/statistical_review_of_world_energy_full_report_2011.pdf (status: February 22, 2013).

Burki, A. A., and D. Terrell, "Measuring Production Efficiency of Small Firms in Pakistan," in: *World Development*, 26 (1998), 155–169.

Busse, Matthias, and Carsten Hefeker, "Political Risk, Institutions and Foreign Direct Investment," in: *European Journal of Political Economy*, 23 (2007), 397–415.

Chang, Sheng-Kai, *Simulation Estimation of Dynamic Panel Tobit Models*, Working Paper, 2004, available at www3.nd.edu/~meg/MEG2004/Chang-Sheng-Kai.pdf (status: March 25, 2013).

Christensen, Jonas Gade, "Democracy and Expropriations," in: *University of Bergen, Department of Economics Working Papers in Economics*, 6 (2011), available at www.uib.no/filearchive/wp06.11.pdf (status: February 22, 2013).

Click, Reid W., and Robert J. Weiner, "Resource Nationalism Meets the Market: Political Risk and the Value of Petroleum Reserves," in: *Journal of International Business Studies*, 41 (2010), 783–803.

Creswell, John W., and Vicki L. Plano Clark, *Designing and Conducting Mixed Methods Research* (Wiley Online Library, 2007).

Crystal, Jill, *Oil and Politics in the Gulf: Rulers and Merchants in Kuwait and Qatar* (Cambridge University Press, Cambridge, 1995).

Eller, Stacy L., Peter R. Hartley, and Kenneth B. Medlock, "Empirical Evidence on the Operational Efficiency of National Oil Companies," in: *Empirical Economics*, 40 (2011), 623–643.

Elsas, Ralf, and David Florysiak, *Dynamic Capital Structure Adjustment and the Impact of Fractional Dependent Variables*, SSRN 1632362 (2010), available at http://papers.ssrn.com/sol3/papers.cfm?abstract_id=1632362 (status: February 22, 2013).

—, *Dynamic Capital Structure Adjustment and the Impact of Fractional Dependent Variables*, Revised version, SSRN 1632362 (2012).

Farzanegan, Mohhamad Reza, "Resource Wealth and Entrepreneurship: A Blessing or a Curse?", in: *MAGKS Papers on Economics*, 201224 (2012), available at http://ideas.repec.org/p/mar/magkse/201224.html (status: February 22, 2013).

Fearon, James D., and David D. Laitin, "Ethnicity, Insurgency, and Civil War," in: *American Political Science Review*, 97 (2003), 75–90.

Griffin, James M., and Weiwen Xiong, "The Incentive to Cheat: An Empirical Analysis of OPEC," in: *Journal of Law and Economics*, 40/2 (1997), 289–316.

Gupta, Sanjeev, Luiz De Mello, and Raju Sharan, "Corruption and Military Spending," in: *European Journal of Political Economy*, 17 (2001), 749–777.

Guriev, Sergei, Anton Kolotilin, and Konstantin Sonin, *Determinants of Nationalization in the Oil Sector: A Theory and Evidence from Panel Data*, Working Paper, 2009, available at www.dsg.fohmics.net/Portals/Pdfs/Kolotilin.pdf (status: February 22, 2013).

—, "Determinants of Nationalization in the Oil Sector: A Theory and Evidence from Panel Data," in: *Journal of Law, Economics, and Organization*, 27 (2011), 301–323.

Güven, Ali Burak, "Reforming Sticky Institutions: Persistence and Change in Turkish Agriculture," in: *Studies in Comparative International Development*, 44/2 (2009), 162–187.

Hamilton, W. Allan, "Sulphate-Reducing Bacteria and the Offshore Oil Industry," in: *Trends in Biotechnology*, 1 (1983), 36–40.

Henisz, Witold J., *POLCON 2005 Codebook*, Manuscript, University of Pennsylvania, 2005.

Herb, Michael, "A Nation of Bureaucrats: Political Participation and Economic Diversification in Kuwait and the United Arab Emirates," in: *International Journal of Middle East Studies*, 41 (2009), 375–395.

Janssen, Marco A., and Bert de Vries, "Climate Change Policy Targets and the Role of Technological Change," in: *Climatic Change*, 46 (2000), 1–28.

Jeong, Yujin, and Robert J. Weiner, "Who Bribes? Evidence from the United Nations' Oil-for-Food Program," in: *Strategic Management Journal*, 33/12 (2012), 1363–1383.

John, Kose, Tae-Nyun Kim, and Darius Palia, *Heterogeneous Speeds of Adjustment in Target Capital Structure*, SSRN 2024357 (2012), available at http://papers.ssrn.com/sol3/papers.cfm?abstract_id=2024357 (status: March 25, 2013).

Johnson, R. Burke, Anthony J. Onwuegbuzie, and Lisa A. Turner, "Toward a Definition of Mixed Methods Research," in: *Journal of Mixed Methods Research*, 1 (2007), 112–133.

Jojarth, Christine, "The End of Easy Oil: Estimating Average Production Costs for Oil Fields around the World," in: *Program on Energy and Sustainable Development Working Papers*, 72 (2008), available at http://cddrl.stanford.edu/publications/the_end_of_easy_oil_estimating_average_production_costs_for_oil_fields_around_the_world (status: February 22, 2013).

Jones Luong, Pauline, and Erika Weinthal, *Oil Is Not a Curse: Ownership Structure and Institutions in Soviet Successor States* (Cambridge University Press, Cambridge, 2010).

Kaiser, M. J., "A Survey of Drilling Cost and Complexity Estimation Models," in: *International Journal of Petroleum Science and Technology*, 1 (2007), 1–22.

Keele, Luke, and Nathan J. Kelly, "Dynamic Models for Dynamic Theories: The Ins and Outs of Lagged Dependent Variables," in: *Political Analysis*, 14 (2006), 186–205.

Leite, da Cunha Carlos, and Jens Weidmann, "Does Mother Nature Corrupt? Natural Resources, Corruption, and Economic Growth," in: *IMF Working Papers*, 85 (1999).

Le Leuch, Honore, and Jean Masseron, "Economic Aspects of Offshore Hydrocarbon Exploration and Production," in: *Ocean Management*, 1 (1973), 287–325.

Lewis, John, "Fiscal Policy in Central and Eastern Europe with Real Time Data: Cyclicality, Inertia and the Role of EU Accession," in: *Applied Economics*, 45 (2009), 3347–3359.

Lledó, Victor, and Marcos Poplawski-Ribeiro, "Fiscal Policy Implementation in Sub-Saharan Africa," *IMF Working Papers*, 172 (2011), available at www.imf.org/external/pubs/ft/wp/2011/wp11172.pdf (status: February 22, 2013).

Loudermilk, Margaret S., "Estimation of Fractional Dependent Variables in Dynamic Panel Data Models with an Application to Firm Dividend Policy," in: *Journal of Business & Economic Statistics*, 25/4 (2007), 462–472.

Luciani, Giacomo, "Allocation vs. Production States: A Theoretical Framework," in: *The Arab State* (University of California Press, Berkeley, CA, 1990), 65–84.

Maddala, Gangadharrao Soundalyarao, *Limited-Dependent and Qualitative Variables in Econometrics* (Cambridge University Press, Cambridge, 1983).

Mahoney, James, "Path Dependence in Historical Sociology," in: *Theory and Society*, 29/4 (2000), 507–548.

Marcel, Valerie, *Oil Titans: National Oil Companies in the Middle East* (Brookings Institution Press, Washington DC, 2006).

Nolan, Peter A., and Mark C. Thurber, "On the State's Choice of Oil Company: Risk Management and the Frontier of the Petroleum Industry," in: *Oil and Governance. State-Owned Enterprises and the World Energy Supply* (Cambridge University Press, New York, 2012), 121–170.

Palacios, Luisa, "The Petroleum Sector in Latin America: Reforming the Crown Jewels," in: *SciencePo CERI Studies*, 88 (2002), available at www.sciencespo.fr/ceri/en/content/petroleum-sector-latin-america-reforming-crown-jewels (status: February 22, 2013).

Papke, Leslie E., and Jeffrey M. Wooldridge, "Panel Data Methods for Fractional Response Variables with an Application to Test Pass Rates," in: *Journal of Econometrics*, 145/1 (2008), 121–133.

Pierson, Paul, "Increasing Returns, Path Dependence, and the Study of Politics," in: *American Political Science Review*, 94/2 (2000), 251–267.

PIW – Petroleum Intelligence Weekly, *PIW Top 50* (Energy Intelligence, December 2011).

La Porta, Rafael, Florencio Lopez de Silanes, Andrei Shleifer, and Robert W. Vishny, "Law and Finance," in: *Journal of Political Economy*, 106/6 (1998), 1113–1155.

Ross, Michael, "Does Oil Hinder Democracy?" in: *World Politics*, 53 (2001), 325–361.

—, "A Closer Look at Oil, Diamonds, and Civil War," in: *Annual Review of Political Science*, 9 (2006), 265–300.

—, *The Oil Curse: How Petroleum Wealth Shapes the Development of Nations* (Princeton University Press, Princeton, NJ, 2012).

Sachs, J. D, and A. M. Warner, "Natural Resource Abundance and Economic Growth," in: *NBER Working Papers*, 5398 (1995), available at www.nber.org/papers/w5398 (status: February 22, 2013).

Stata 12 Manual, *Data Analysis and Statistical Software*, available at www.stata.com/manuals/(status: February 28, 2013).

Stroebel, Johannes, and Arthur van Benthem, "Resource Extraction Contracts under Threat of Expropriation: Theory and Evidence," in: *USAEE Working Papers*, Cleveland, 2010.

Tashakkori, Abbas, and Charles Teddlie, *Handbook of Mixed Methods in Social & Behavioral Research* (Sage Publications, London, 2002).

Tavares, J., "Does Foreign Aid Corrupt?" in: *Economics Letters*, 79 (2003), 99–106.

Tétreault, M. A., "The Political Economy of Middle Eastern Oil," in: *Understanding the Contemporary Middle East* (Lynne Rienner, Boulder, CO, 2008), 255–279.

Thelen, Kathleen, "How Institutions Evolve: Insights from Comparative Historical Analysis," in: *Comparative Historical Analysis in the Social Sciences* (Cambridge University Press, Cambridge, 2003), 208–240.

Tilly, Charles, *The Politics of Collective Violence* (Cambridge University Press, 2003).

Tordo, Silvana, Brandon S. Tracy, and Noora Arfaa, *National Oil Companies and Value Creation* (World Bank Publications, Washington DC, 2011).

Transparency International, *Promoting Revenue Transparency: 2011 Report on Oil and Gas Companies*, March 2011, available at www.transparency.org/whatwedo/pub/promoting_revenue_transparency_2011_report_on_oil_and_gas_companies (status: February 28, 2013).

Tsui, Kevin, "Resource Curse, Political Entry, and Deadweight Costs," in: *Economics & Politics*, 22 (2010), 471–497.

Vaaler, Paul M., and Burkhard N. Schrage, "Residual State Ownership, Policy Stability and Financial Performance Following Strategic Decisions by Privatizing Telecoms," in: *Journal of International Business Studies*, 40 (2009), 621–641.

Victor, Nadejda, "On Measuring the Performance of National Oil Companies (NOCs)," in: *PESD Stanford University Working Papers*, 64 (2007), available at http://pesd.stanford.edu/publications/nocperformance (status: February 22, 2013).

Warshaw, Christopher, "The political economy of expropriation and privatization in the oil sector," in: *Oil and Governance. State-Owned Enterprises and the World Energy Supply* (Cambridge University Press, New York, 2012), 35–61.

Wolf, Christian, "Does Ownership Matter? The Performance and Efficiency of State Oil vs. Private Oil (1987–2006)," in: *EPRG Working Paper*, 813 (2008), available at www.eprg.group.cam.ac.uk/wp-content/uploads/2008/11/eprg0813.pdf (status: February 22, 2013).

—, "Does Ownership Matter? The Performance and Efficiency of State Oil vs. Private Oil (1987–2006)," in: *Journal of Energy Policy*, 37 (2009), 2642–2652.

Wooldridge, Jeffrey M., *Econometric Analysis of Cross Section and Panel Data* (MIT Press, Cambridge, MA, 2001).

—, *Introductory Econometrics: A Modern Approach* (South-Western Publishers, Nashville, TN, 2009).

World Bank, *Deterring Corruption and Improving Governance in the Electricity Sector* (World Bank Publications, Energy, Transport & Water Department and Finance, Economics & Urban Department, Washington DC, 2009).

Data sources – statistical analysis

British Petroleum (BP), *Statistical Review of World Energy 2011*.

Energy Information Administration (EIA), *World Crude Oil Production (1970–2009)*.

Henisz, Witold, *Polcon III Dataset*, 2010.

International Monetary Fund (IMF), *World Economic Outlook 2012*.

Oil and Gas Journal (OG & J), *Historical Worldwide Oil Field Production Survey (1980–2007)*.

Petroleum Intelligence Weekly (PIW), *PIW Top 50 (1987–2010)*.

Political Risk Services (PRS), *International Country Risk Guide data (1984–2010)*.

World Bank, *Education Statistics Database*.

World Bank, *World Development Indicators*.

5 Mirroring the cases of Saudi Arabia and Abu Dhabi

The quantitative analysis has uncovered that the context shaped by geology, technical know-how of the National Oil Companies (NOC), and international oil prices counts only partly for the upstream sector policy pursued by oil producing countries. It is the extent of national oil reliance and executive constraints which shape the oil upstream sector policy adopted by state leaders and the executive given their primary aim of political survival. To better understand the relevance of these driving forces at play, two case studies in the following section—on Saudi Arabia and Abu Dhabi in the United Arab Emirates (UAE)—are to offer closer insights into the functioning of a closed and respectively, open oil upstream sector.

In reference to Seawright and Gerring, a case study needs to be conceived as "the intensive analysis of a single unit or a small number of units (the cases), where the researcher's goal is to understand a larger class of similar units (a population of cases)" (2008, 296). The problem of inference from a sample to a larger population is inherent (Gerring 2004, 342, and 2007; Seawright and Gerring 2008, 296), yet the careful case selection—discussed in section 5.1 below—and the advantages of a mixed method design (coupling statistical analysis with case studies) should strengthen the results.

The primary data collected through semi-structured expert interviews—to be introduced in section 5.2—and analyzed in the framework of the case studies—sections 5.3 and 5.4—will be confronted, where existent, with secondary sources. Notably, the literature on the political economy of Saudi Arabia and more specifically, its petroleum sector is quite rich, at least as compared to the works on Abu Dhabi, which are extremely scarce. From this vantage point, this book not only draws on secondary data but rather contributes to the empirical scholarship on Saudi Arabia and Abu Dhabi/the UAE with primary information. The case studies begin with some historical background based on secondary literature and continue with the discussion of the analytical framework in reliance on primary data cross-checked with secondary literature, where existent.

5.1 Case selection

Saudi Arabia and Abu Dhabi build the focus of the case studies. Both are located in the Middle East, which is not only the most oil-endowed region in the world but it also has the largest oil output per day (British Petroleum 2011). Given the relevance of the Middle East for the global oil market, it is imperative to understand how the region sheds light on the research question of this book. Among the Middle Eastern oil producing countries, the preference for Saudi Arabia and the UAE with the emirate of Abu Dhabi is informed by several reasons.

First, the quantitative analysis has pointed to a number of trends across the sample and largely validated the relations outlined in the analytical framework. Statistically, the interaction between low executive constraints and high national oil reliance is likely to lead to more state control in the oil upstream sector. Both Saudi Arabia and Abu Dhabi have low political constraints and high national oil reliance, yet their oil upstream sector policies lie at the extreme—the former with a closed upstream sector (as predicted in the statistical analysis), the latter with an open upstream sector where state control has never topped 100 percent. For a graphical representation of the control structures in Saudi Arabia versus Abu Dhabi in the oil upstream industry in the timeframe 1987 to 2010, please refer to Annex 15.

Second, not only do Saudi Arabia and Abu Dhabi have widely divergent outcomes on the dependent variable, but a plot of state control by countries (see Annex 16) also shows that they have pursued largely consistent policies across time. Among the Middle East and North African (MENA) oil producing countries in the sample, a similar pattern to that pursued by Saudi Arabia, i.e. state control ratio close to 1, can be identified in the cases of Iran, Iraq and Kuwait. As for the open upstream policy pursued by Abu Dhabi/the UAE,[1] a relatively similar trend is discerned for Egypt and Oman. Yet, across each group of MENA producer countries with constantly closed and respectively, open upstream sectors, the preference has gone for the country with the largest oil reserves as these are more likely to set the tone on the international oil market and raise more attention as to the global energy security agenda. Saudi Arabia has the world's largest oil reserves (BP est. 2011: 265 billion barrels) whereas Abu Dhabi/the UAE has the fifth largest oil reserves globally (BP est. 2011: 97 billion barrels), considerably larger than Egypt or Oman.

While in absolute terms, Saudi Arabia has about two and a half times more oil reserves than Abu Dhabi, these figures need to be linked to demographics. Saudi Arabia, with a population of approximately 27 million people, has a ratio of approximately 10,000 barrels oil/capita whereas Abu Dhabi, with a population as small as 1,600,000 people, has a ratio of 58,000 barrels oil/capita.[2] In other words, the oil reserves endowment is so large in both cases that they are—beyond discussion—highly relevant not just for the oil production in the Middle East region but worldwide.

Last but not least, data accessibility also spoke for the selection of the two cases. Thirty-three energy experts on Saudi Arabia and Abu Dhabi/the UAE were accessible and agreed to interview for the purpose of this book publication in Washington DC and the area.[3] The interviewing technique will be discussed more narrowly in the next section.

5.2 Primary data collection through semi-structured interviews

Semi-structured interviews represent the primary data collection method applied for the case studies in this book. The interviewing technique using a semi-structured format allows both closed and open questions and enables a conversational style (Legard *et al.* 2003; Millwood and Heath 2008). There are several advantages to this data collection method which has proven fruitful for the present research. First, while the interviewer creates a questionnaire with the topics in advance, the order of questions is not fixed as the semi-structured interviewing technique allows the interviewee to follow her train of thought. This enables access to new information of which the interviewer might not have been aware at the very start. Second, the open questions don't just stimulate discussion and prevent omissions but also guide the interviewee lest she might lose herself in digressions (always an imminent danger). Third, the language and explanations of questions can be easily adapted to the individual interviewee given her jargon and expertise in the field—whether in the industry, academia, think-tank, consulting and so on (Dexter 2006; Mason 2002).

For the purpose of this research, the baseline questionnaire comprised 23 both closed and open questions and was structured in four parts. In the first part, the aim was to map out the organization of the petroleum sector, identify the main actors and determine the role of the state and the NOC in the oil industry. The second part looked into the role of international oil companies (IOCs) while addressing their mandate in the upstream sector and the respective country, and understanding the requirements for control rights. The third section focused on the benefits and potential pitfalls of the oil upstream policy pursued in the respective country. Fourth and finally, the relevance of the petroleum industry to the national economy was addressed and the potential for sustainable development in the oil producing country was inquired. Depending on the elaboration of answers to the open questions, the interview took from 40 to 90 minutes. For the full questionnaire, please refer to Annex 17.

Already in the first couple of months (September/October 2011), when contacts with potential interviewees were established, the sensitivity of the topic became apparent. It turned out to be not just the oil-related side but also the regional focus which made it more uncomfortable for high officials, established industry professionals as well as think-tank researchers to express views on the upstream investment regimes of the two oil producing countries. Consequently, for the sake of access to key experts and reliable data, confidentiality was guaranteed to all interviewees. In his interviews on

Petromin and the development of the Saudi oil sector, Steffen Hertog faced the same difficulty and emphasized that: "[w]ithout this [anonymity], I would effectively have had no access to the sensitive information and frank assessments that my interviewees shared with me and that I am deeply grateful for. Anonymity of sources is a price one often still has to pay for researching Saudi politics" (Hertog 2008, 647).

For the present research, 33 experts were interviewed in total. Regarding the definition of an "expert," the scholarship on the methodological foundations of the "expert interview" takes three different approaches to the notion (Bogner and Menz 2005, 48ff.). Despite opening itself to various criticisms in particular from the sociology of knowledge (Bogner and Menz 2005, 50ff.), this book relies on the method-relational approach, which assumes that the "expert" is in possession of relevant knowledge about a certain subject in her quality as a trained and specialized professional (ibid., 49f.) or as a member of the "functional elite" (Meuser and Nagel 1994, quoted in Bogner and Menz 2005, 50). More explicitly, only those individuals are considered experts who possess special knowledge accumulated as a "socially institutionalized expertise," which is widely linked to the role of the professional (Sprondel 1979, quoted in Meuser and Nagel 2009, 19).

Therefore, experts were considered to be academic scholars doing research on the economy of Saudi Arabia and/or Abu Dhabi in the UAE, think-tank experts and consultants with the respective area of expertise, oil industry professionals, government officials and international organization diplomats focusing on energy in the MENA region. Most of the interviews were conducted at high and very high level—for example, with one of the former Chairmen of Mobil/ExxonMobil in Saudi Arabia, former U.S. Ambassadors to Saudi Arabia, senior economists of Saudi Aramco, heads of MENA departments in international organizations, current vice-president of one of the major oil companies worldwide, directors in the U.S. State Department and the U.S. Department of Energy, well-established researchers and professors from first-tier universities such as Georgetown University, London School of Economics, Cambridge University and others.

Different perspectives stemming from different viewpoints were thus raised. In reference to the professional field, the distribution of the 33 interviewees looks as follows:

- Energy consulting: three (IHS Cambridge Energy Research Associates (IHS CERA), Deutsche Bank, and Manaar Energy Consulting).
- Industry: four (Saudi Aramco, ExxonMobil, and Statoil).
- International organizations (IOs): four (World Bank and IMF).
- Government: six (U.S. Department of Energy and U.S. Department of State—both current and retired employees).
- Think-tanks: six (Centre for Strategic and International Studies (CSIS), Brookings, Carnegie, King Abdullah Petroleum Studies and Research Centre (KAPSARC), and Chatham House).

- Academia: ten (Georgetown University, Brown University, Cambridge University, London School of Economics, UAE University, Dubai School of Government, Graduate Institute in Geneva, and University of Texas at Austin).

Notably, some of the interviewees who were categorized under academia have also been engaged with energy consulting. They either own their own consulting companies or are employed as independent consultants by the governments of Saudi Arabia or Abu Dhabi/the UAE. The same is the case with some of the former government officials interviewed for this research.

Three of the interviews were collective, namely with two and respectively, three interviewees simultaneously. Moreover, five experts were interviewed via Skype as they were either in the Middle East or in Europe at the time and no prospects of meeting them within six months were envisaged. Three other experts were willing to reply to questions via email, whereas all the other 25 experts were interviewed in person. The interviewing phase as such, including establishing contacts through chain referral or the so-called "snowball sampling" technique,[4] took eight months in total. The location, Washington DC, was strategic as it proved to be the right venue to meet experts from all over the world.

One clear observation during the interviewing phase was that the community of experts for the UAE oil sector is considerably smaller than that for Saudi Arabia. Consequently, 12 out of the 33 interviewees are experts specifically on Saudi Arabia whereas only six on Abu Dhabi. Ten interviewees were confident to answer questions on both the Kingdom's and the Emirate's oil sectors. The rest of the interviewees (five) helped to clarify some of the technical questions related to geology and technological/technical requirements, as well as legal issues on upstream contracts and IOCs–NOCs consortia. Consequently, the professional distribution of the 22 interviewees on Saudi Arabia and 16 interviewees on Abu Dhabi looks as follows:

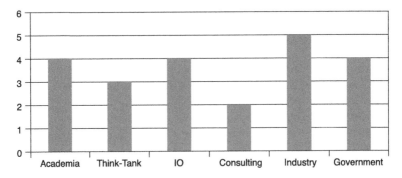

Figure 5.1 Professional distribution of interviewees on Saudi Arabia (22 interviewees)

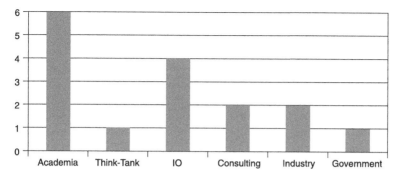

Figure 5.2 Professional distribution of interviewees on Abu Dhabi/the UAE (16 interviewees)

Due to the issue of data sensitivity, apart from the three interviewees who responded in writing through email, only 19 of the other 30 interviewees agreed to be recorded. For the remaining 11 interviewees, notes were taken during the interview, all of them conducted in person.

Lest the quotes or paraphrases be attributed to one specific individual, the case studies below will not name the interviewee's explicit institutional affiliation, with the exception of Saudi Aramco. This was one of the collective interviews where the interviewees gave the official position of the institution and allowed the quotes and information to be explicitly associated with Saudi Aramco.[5] As for all the other interviewees, the broader field (i.e. consulting, industry, IOs, government, think-tank or academia) will be mentioned together with the date of the interview for differentiation purposes (e.g. academia, October 28, 2011; industry, December 13, 2011; think-tank, January 26, 2012). This way, the pledge of anonymity can be kept and at the same time, the views from different sectors on the oil production policies of Saudi Arabia and Abu Dhabi can be delineated.

5.3 Closed upstream sector: Saudi Arabia

The birth of the oil industry in Saudi Arabia coincides with the birth of the modern Saudi state. A brief historical account thereof is not just in order but quite imperative for the intricacies of the oil production policy in Saudi Arabia to be understood. The chapter will then continue with the discussion of the proposed analytical framework based on the interviews and when available, secondary literature.

5.3.1 Historical contingencies

Abd al-Aziz ibn Saud was the founder of the third Saudi state or the Kingdom of Saudi Arabia.[6] The creation of the modern state commenced in 1901 when Abd al-Aziz with a group of 40 men set out from Kuwait to recapture

Riyadh, which had been under the rule of the Al Saud—yet intermittently—since the middle of the eighteenth century.[7] A heartfelt adherent of the Muwahhidin (i.e. monotheist) revival, Abd al-Aziz raised the help of the Ikhwan or "the Brethren," "fanatical, ascetic Islamic warriors" (Long and Maisel 2010, 34), to conquer back former Saudi territories. In 1926, he had already established his rule over the Hijaz and Najd, including its Dependencies, while Asir was annexed three years later. International recognition of Abd al-Aziz's new rule came with the Treaty of Jiddah in 1927.

By 1932, Saudi Arabia was a "desert kingdom" (ibid.) with poverty as the main obstacle to development. Before oil, the main sources of income were pastoral agriculture in the interior, trade (pearl fisheries, export of Arabian horses to India, agricultural products from Asir and al-Hasa) and the hajj (levies on pilgrimage to Makkah and Madinah). As Niblock and Malik point out, "[t]he social characteristics which accompanied this form of economic life were those of relative poverty, a high level of illiteracy, and substantial isolation from developments outside of the local area" (2007, 35).

The turning point in the economy of Saudi Arabia came with the signing of the first oil concession with Standard Oil of California (SOCAL) in 1933 for a period of 60 years (Myers Jaffe and Elass 2007; Stevens 2012). This marked the beginning of the oil kingdom. The wholly owned subsidiary of SOCAL, i.e. California Arabian Standard Oil Company (CASOC), started drilling in April 1935. By 1938 large quantities of oil were proven commercial. However, due to logistical problems, SOCAL had to look for partners and "in July 1936, SOCAL and Texaco converted CASOC into a fifty-fifty joint venture—they married SOCAL's Saudi supply with Texaco's downstream markets" (Stevens 2012, 176). In 1944 CASOC became the Arabian American Oil Company (Aramco) and four years afterwards, two more partners stepped into the private consortium, as follows: SOCAL (30%), Texas Oil Company (Texaco: 30%), Standard Oil Co. of New Jersey (Exxon: 30%) and Socony-Vacuum (Mobil: 10%). In 1947, Aramco was already producing 90,000 barrels of oil per day (Marcel and Mitchell 2006, 29; Stevens 2012).

Oil exports started in 1948, as a consequence of which governmental revenues soared from US\$ 53.6 million in 1948 to US\$ 100 million in 1953 and US\$ 333.7 million by 1960 (Niblock 1982, 95f.; Niblock and Malik 2007, 36). Rapidly growing oil income created the basis for a differentiated administrative apparatus which was supposed to depart from a system revolving around one single person, Abd al-Aziz, "who was king and chief legislator and had the right to review all judicial decisions" (Hertog 2007, 542). In the absence of independent powerful tribes and also, of a strong-minded "ulama," the institutional design of the modern state was shaped by the Al Saud family structure. While the tribes were crushed in the 1920s and 1930s during Abd al-Aziz's take-over of the peninsula, the Saudi "ulama" raised no constraints on the royal family as it has been constantly guided by pragmatism with respect to Al Saud's prerogatives (Steinberg 2005). In Hertog's opinion, "Saudi bureaucracy building often seems not to be a case of 'form follows function' but of 'form follows family'" (2007, 555).

The state apparatus grew to become a behemoth in the 1960s/1970s. The origins of the cooptation regime lie in those formative years which were decisive for the structure of modern institutions. Before his death in late 1953, for the first time in the history of Saudi Arabia, Abd al-Aziz organized the heads of the ministries/other agencies into a cabinet which met in 1954 under the new King Saud (1953–1964). The composition of the Saudi government clearly pointed to a dynastic monarchy, where the highest state offices or what Herb calls "ministries of sovereignty" (1999, 8) were occupied by the sons or grandsons of Abd al-Aziz ibn Saud, i.e. the founder of the Kingdom[8] (ibid., 92f.). King Saud's mismanagement of the oil reserves in the 1950s (Long and Maisel 2010, 41) gradually led to the centralization of power in the hand of Prince Faisal, initially Prime Minister (1958–1960, and 1962–1964) and afterwards King (1964–1975) (Niblock and Malik 2007, 51). The inflow of oil revenues set "the lineages of the rentier state" (Chaudhry 1994), whereby the ruling factions of the Al Saud enjoyed both "structural autonomy" (i.e. no need to tax) and "political autonomy" (i.e. unrestricted policy-making) at the beginning of the Saudi state-building process (Chaudhry 1989, 116ff.). Furthermore, "oil rents created a huge, financially autonomous distributive bureaucracy" and blocked the development of both regulatory and extractive institutions in Saudi Arabia (ibid., 103).

There were four main channels through which wealth was distributed from the royal hub to the population during the boom years (1973–1982). These were: (1) subsidies for basic goods and utilities, free higher education, welfare and social security programmes for the society at large; (2) construction, farming and industrial subsidies for land-owners, also known as urban and rural land grants, for which personal and informal connections played an immense role; (3) large interest-free loans for housing, industrial and agricultural development, personal needs etc. in particular to Nejd to the detriment of other regions (i.e. 60–80 percent of the projects allocated to 10 percent of the population); and (4) commissions for labour import and the establishment of new enterprises, which were obtained primarily by those with access to the Al Saud or other high government officials (ibid., 126f.).

Through these spending patterns, the allocative Kingdom of Saudi Arabia generated a new social stratification while dismantling existing occupational groups and destroying their social status—such as of the former merchants (Field 1984). Based on close kinship and patrimonial links to the bureaucracy, the newly state-created business class then managed to oppose the state efforts for economic reform in the bust years of the 1980s (Chaudhry 1994; Okruhlik 1999). To this day, the effect of the formative years in the history of the modern Saudi state has been deeply felt on both state capacity and business–government relations (Chaudhry 1994; Hertog 2007, 2010a). To Luciani, "the state remains, as it has always been, the protector of the business community—from the origins of the Saudi state itself" (2005, 157).

Nonetheless, "princely networks of (re)distribution have not been allowed free rein everywhere" so that institutional "islands of efficiency in the state" have emerged and have been kept insulated from games of patrimonial politics and brokerage (Hertog 2010a, 28 and 2010b). The efficiency of some state agencies like the Central Bank SAMA (Saudi Arabian Monetary Agency), the petrochemicals giant SABIC (Saudi Basic Industries Corporation), the Saudi Ports Authority and the NOC Saudi Aramco has become a fact (Hertog 2008, 2010b; Kobayashi 2007; Luciani 2005; Seznec and Kirk 2010; Stevens 2012; *The Economist*, January 21, 2012). In reference to Hertog's extensive writings on the subject, this is beholden to a few senior Al Saud princes.

Moreover, the "state-led development policy" (Niblock and Malik 2007, 51) has not been confined to the economic landscape but has extended to the political and societal domains. Out of "defensive modernization," the Saudi echelons have tried to formally engage some groups in political decision-making and feigned interest in consultation and dialogue. In this sense, since the beginning of the 1990s several institutions have been created: the Consultative Council (Majlis al-Shura), the National Dialogue and the organization of municipal elections (Ammoun 2006). "Old wine in new bottles," these institutions are not guided by the logic of representation but still of co-optation. That is, "[t]he Majlis al-Shura, rather than representing Saudi society, represents the royal family's different clans, their political and economic interests, their ideological and political tendencies, and also, more importantly (and through them), their social bases, meaning their clientele" (ibid., 225).

Despite some efforts at modernizing the Saudi politics, no actual steps have been taken towards democratization. Kinship, friendship and patronage relations remain the core (informal) structures of political interest articulation in an absolutist monarchy (Herb 2004) or if you will, a "liberalized autocracy" (Brumberg 2002). If any, political mobilization in Saudi Arabia in the past 20 years has come from above (Ehteshami and Murphy 1996; Ehteshami and Wright 2007; Hertog 2004, 2006).

As late as the twenty-first century, Saudi Arabia continues to be run as a "family business" (Halliday 2000) and to remain a state of personalities rather than institutions. Madawi Al-Rasheed (2005) distinguishes five main circles of power, small states within the state, revolving around royal princes, their sons and a wide network of commoners. At Al-Rasheed's moment of writing, these factions were those of King Fahd (deceased in 2005), Prince Nayef (Minister of the Interior, died in 2012), Crown Prince Abdullah (also Commander of the Saudi Arabian National Guard, since 2005 King of Saudi Arabia), Prince Sultan (Minister of Defence, since 2005 Crown Prince, died in 2011), and Prince Salman (Governor of Riyadh, since 2011 Minister of Defense and since 2012 Crown Prince). Apart from these ruling factions, there are a number of other aspiring princely circles such as those of Al Faisal, Abd al-Rahman, Turki and Ahmad (Al Rasheed 2005, 200–208). A fiefdom of a powerful prince or family branch can still go as broad as to encompass "a bureaucratic (ministry), a security (intelligence service, military or SANG), a social (hospitals, welfare services or foundations) and/or an intellectual domain (research centers, libraries and newspapers)" (Glosemeyer 2005, 218; same idea in Hertog 2005, 119).

All in all, the aim of this historical record was to provide a background on the institutional structures which have been forged since the discovery of oil and which can be still seen in the current setup of Saudi Arabia. Despite the stability of the past five decades, social and economic changes might impact the equilibrium heretofore. By comparison with the 1960s, the population has increased by over six times, from about 4 million to over 27 million nowadays, based on the CIA World Factbook 2012. In 2010, 52 percent of the Saudi population was under the age of 25 (Salehi-Isfahani 2010). The same year, the unemployment rate of Saudi youth (at approximately 30 percent) was almost three times as large as that of Saudi nationals, while foreign workers (4.7 million) continued to outnumber the nationals (4.3 million) in the labour force (ILO 2011). In the face of these demographic trends, the domestic consumption of oil and gas has been growing very sharply, which might eat out of the amount of oil for export. In this sense, Jadwa Investment draws attention to the fact that "[l]ow domestic prices—oil is sold in Saudi Arabia at between 3 percent and 20 percent of the global price—mean the efficiency of oil use is worsening. Consumption is growing at about twice the rate of non-oil GDP growth" (2011, 2).

Despite potential reasons for unrest, especially given the labour market, Saudi Arabia has had no Arab Spring. Proactive measures were taken against instability and any form of dissent was repressed in its infancy (Abu-Nasr 2012; BBC, August 31, 2012; Blight *et al.* 2012). The royal family proved itself skillful in preempting unrest in the country. In 2011 King Abdullah announced "a $130 billion plan to create jobs, build subsidized housing and support the religious establishment that had backed the government's ban on domestic protests" (Abu-Nasr 2012) and promised reforms regarding women's voting and candidacy in elections as of 2015 (BBC, August 31, 2012). These measures themselves point to the fact that tensions are sizzling under the surface and the future might not look just as stable as the past.

5.3.2 The analytical framework against the Saudi Empirics

In the following, the proposed analytical framework is confronted with the Saudi realities. This section is largely based on primary data as well as secondary sources. In reference to the structure of the analytical framework, the domestic and international context of the oil upstream sector policy in Saudi Arabia is first mapped out. The focus will then shift to the domestic constraints of politico-institutional and economic nature.

5.3.2.1 The context

To start off with the *geological conditions*, Saudi Arabia is clearly more privileged than any other oil producing country in the world. As one of the think-tank experts pointed out, "Saudi Arabia has the easiest oilfields in the world to produce" (Think-tank, October 7, 2011). All 22 interviewees have agreed on the favorable geology of Saudi Arabia, which has about 20 percent of the global oil reserves (see also EIA 2013a and BP 2011). In an article published in OPEC Review, Al-Attar and Alomair provide upstream costs data in a number of selected countries, among which Saudi Arabia has by far the lowest exploration and production costs of $3 per barrel (2005, 250).

Following grave allegations made by Matthew R. Simmons already in the early 2000s, before the publication of his book *Twilight in the Desert: The Coming Saudi Oil Shock and the World Economy* (2005), Saudi Aramco disclosed a lot of heretofore confidential data on both reserves and operations during a presentation at the CSIS in Washington DC on February 24, 2004 (Baqi and Saleri 2004). Based on this, the average total depletion of the Saudi oilfields is around 28–29 percent. The oldest oilfield, Abqaiq, is 73–74 percent depleted, the giant Ghawar (the largest oilfield globally) has produced 48 percent of its proven reserves, while the Shaybah just 5 percent. This data was reconfirmed by Saudi Aramco during the interview and it is also publicly available in the EIA Country Analysis Brief on Saudi Arabia (2013a). While the annual depletion rates of the oilfields operated by IOCs range between 4.2 percent (Prudhoe Bay) and 9.6 percent (Brent), Saudi Aramco follows a clear depletion policy of 2 percent per year (industry, March 25, 2012; see also Baqi and Saleri 2004, slide 14). This is highly relevant for the very essence of Saudi Aramco and the organization of the oil upstream sector in Saudi Arabia.

More specifically, while IOCs are first and foremost concerned with access to oil reserves, Saudi Aramco is likely to be producing from the Saudi oilfields for at least 50 years. Therefore, the access to reserves represents no difficulty. Yet the concerns of the Saudi oil company are of a different nature, as Saudi Aramco has to deliver on more than just purely commercial objectives. First, as shown above, the company does not produce oilfields as quickly as private companies do because the Saudi government has decided that the fields should maximize production of the oil in place and not necessarily revenues. "The decision was made in Riyadh—so we deplete our fields at 2% a year whereas a regular company runs through the cycle within 15 years," stated one of the interviewees at Saudi Aramco (December 13, 2011). Second, no commercial oil company wants spare capacity. Yet following the policy set by the Ministry of Petroleum and Mineral Resources, Saudi Aramco has not just the world's largest crude oil production capacity of 12 million barrels/day but also upholds spare capacity of 2 million barrels/day (EIA 2013a). This production scheme is in line with the "swing producer" status assumed by Saudi Arabia on the international oil market and it is explicitly disposed by the Saudi government "to be able to react to market disruptions at a rapid pace" (Saudi Aramco, December 13, 2011). Both depletion rate and spare capacity, as externalities from being a state-owned oil company, were mentioned by seven other interviewees from international organizations (January 24, 2012 and February 15, 2012), think-tank (October 7, 2012), industry (March 15, 2012) and academia (October 28, 2011). The trade-off between the national mission and the long-term accessibility of reserves is apparent. On the one hand, Saudi Aramco follows clear strategic objectives outlined by the Saudi Ministry of Petroleum and Mineral Resources in its quality of exclusive shareholder. On the other hand, the company is in the position to take a long-term approach to operations given its indisputable access to reserves both in the present and the more distant future.

An example of the benefits emerging from the long-term approach is Shaybah. The oilfield was found in the late 1960s but production only started on July 1, 1998. The reasons for this are the location of Shaybah, the difficult topography and climate, which made production unfeasible given the technology status of the late 1960s. Thanks to its rich reserves as well as the operational freedom to decide over the system on the ground, Saudi Aramco "had the luxury to put Shaybah aside for a while and let the technology catch up. So, it stayed in the stockroom for 30 years. Once the horizontal and multilateral drilling with GPS has been developed, it became more cost-effective to put the oil pipeline" (Saudi Aramco, December 13, 2011). Notably, it was not because of a shortage of oil that Saudi Aramco went back to Shaybah but instead, the oilfield became feasible and economically efficient to produce. Sitting on a huge oilfield and postponing production is a privilege that a private company can seldom enjoy because of restricted access to reserves. As a single operator in the Saudi oil upstream sector, Saudi Aramco was able to optimize the decisions across the entire system given the geological conditions, high-tech requirements and the available technology at the moment (industry, March 15, 2012 and academia, October 28, 2011).

Related to the *technical and technological capabilities of Saudi Aramco*, all 22 interviewees emphasized that Saudi Aramco is comparable to any IOC. Three of the interviewees (consulting, January 26, 2012; think-tank, November 29, 2011; and academia, October 28, 2011) compared it with ExxonMobil. Four of the 22 interviewees (think-tank, October 7, 2011; academia, February 9, 2012; government, February 14, 2012 and industry, March 15, 2012) could not prevent themselves from noticing the exceptionality of the advanced technical and technocratic level of

Saudi Aramco "in a sea of entropy." This resonates with Hertog (2010a)'s approach to "islands" or "pockets of efficiency" in Saudi Arabia.

It is a fact that Saudi Aramco has invested a lot in research and development (Marcel 2006; Stevens 2012) and it is particularly involved in reservoir management of the Saudi oilfields for the purpose of maximum hydrocarbon recovery. This has been reaffirmed in the interview with Saudi Aramco in a different form:

> The specific needs of the company drive the kind of R & D and the technological innovation. Those fields such as Ghawar are uniquely structured—not just in size but in structure. We have the absolute best upstream technology in the world for those fields.
>
> (December 13, 2011)

In the case of singular operations involving high level of complexity, Saudi Aramco contracts service operators like Schlumberger, Baker Hughes or Halliburton. In reference to experts from the industry (December 8, 2011), think-tanks (October 7, 2011 and January 26, 2012), academia (October 28, 2011 and March 7, 2012) and government (January 24, 2012 and February 1, 2012), this has been the common practice for IOCs over the years as well.

The technical prowess and technological endowment of Saudi Aramco are not standalone but need to be linked to the history of the company. In fact, 17 of the 22 interviewees touched upon or went into long monologues about the origins and evolution of Saudi Aramco as well as the role of personalities in shaping the Saudi oil industry. Considered by many an "accidental NOC" (Marcel 2006; Myers Jaffe and Elass 2007; Stevens 2012), Saudi Aramco is the result of increased participation—that is, gradual purchase of foreign oil assets in Aramco in the 1970s, due to the events and increasing pressure in the region (Gause 2010; Sorenson 2007; Tétreault 2008).

Only three of the 22 interviewees (academia, February 9, 2012; government, February 14, 2012; industry, March 15, 2012) cared to mention Petromin, which was supposed to become the actual NOC of Saudi Arabia. Drawing on Steffen Hertog's historical account, Petromin was created through a royal decree on November 22, 1962 and initially, it was "responsible for all exploration, refining, and distribution of petroleum and mineral resources in the kingdom that were not in the domain of then U.S.-controlled oil concessionaire Aramco" (Hertog 2008, 650). The leadership of the organization was given to Abdulhadi Taher, a protégé of Ahmad Zaki Yamani, who in March 1962 was made Minister of Petroleum and Mineral Affairs by (at the time) Crown Prince Faisal. Yamani replaced Abdullah Tariki, who as the first Minister of Petroleum (1960–1962) under King Saud, already envisioned Aramco's nationalization and was not particularly accepted in Faisal's circle of power. As the national champion of Saudi industrialization in the 1960s and 1970s, Petromin embarked on a large spate of projects with domestic patrons and clients, but also with international partners. While the company took over every possible industry sector (e.g. refining, mining, petrochemicals, fertilizers, iron, steel, aluminum manufacturing, transport and distribution of petroleum products and so forth), most of its projects did not come to fruition. In short, huge investments led to marginally improved economic results.

In the 1970s, given the regional trend towards nationalization, the Saudi government opted for incremental participation in Aramco (Myers Jaffe and Elass 2007; Yergin 1991, 2009). Under the 1972 General Agreement, the Aramco parents concurred in renouncing equity rights but retained managerial control over operations for a rather long time afterwards. A former employee from one of the four shareholders of Aramco, who witnessed the "pseudo take-over," explained that the IOCs remained in Saudi Arabia until the early 1990s, had no ownership rights in the company anymore, but received a "retained royalty" to continue operations. The transfer initiated under King Faisal was very smooth despite the fact that the four shareholders of Aramco were only paid the book value of their assets. The complete purchase of Aramco's assets by the Saudi government was not publicly announced until September 1980 and it took eight more years for "Saudi Aramco" to be created through a royal decree (industry, March 15, 2012).

Aside from one corruption scandal in 1977/1978 on a gas project, which resulted in the resignation of the chairman Frank Jungers, Aramco has always been considered a very trustworthy organization by the Saudi state. This trust that the government has placed in Aramco/Saudi Aramco was emphasized not just by Saudi Aramco (on December 13, 2011) but it was confirmed by four other experts both from the industry (March 15, 2012) and the government (February 1 and 14, 2012). Based on historical records, Hertog contends that "Aramco as an organization had been widely insulated from local politics. Petromin's opacity and its de facto role as slush fund for well-connected players appeared chronic by comparison" (2008, 656). A bureaucratic behemoth, with over 4,000 employees in the mid-1970s, Petromin was the epitome of corruption, sluggishness and patrimonial relations (ibid.).

In the history of Aramco, Ali Al-Naimi became the first Saudi President of Aramco in 1983 while American John Kelberer remained Chairman and CEO. Five years later, Al-Naimi was also offered the CEO position and Hisham Nazer was appointed the first Saudi Chairman of the company. The latter had replaced Zaki Yamani at the apex of the Oil Ministry under King Fahd. "Under the king's and Nazer's supervision, Aramco's structures had been definitely cocooned" (ibid., 659) and the American corporate culture has been preserved in its genuine form.

The period from 1986 to 1996 consolidated the position of Saudi Aramco, which gradually took over Petromin's former turf—i.e. Saudi Arabian Marketing and Refining Company (Samarec), Petrolube and Luberef. In reference to one of the experts, this included mostly real estate rather than actual deals as "it was difficult to unwind all the corrupt deals which Petromin was enmeshed in and Saudi Aramco was careful not to be infected by Petromin" (academia, March 7, 2012). Already by 1993, with the incorporation of Samarec, former Aramco, by then Saudi Aramco, evolved to "what Petromin was originally meant to become: the sole actor in the Saudi oil sector" (Hertog 2008, 660).

As two of the interviewees (both of them from academia, on February 9, 2012 and March 7, 2012) highlighted, it was a crucial decision for the Saudi oil sector to ditch Petromin and keep Aramco. While Petromin was very likely to become an inefficient, corruption-breeding, patronage-led NOC, Aramco had all the assets in store to become a high-flier. In one of the interviewees' strong opinion, "there is nothing like Saudi Aramco in the Gulf. You would have to look for Petronas or Statoil to find something similar" (academia, March 7, 2012). There is more to Saudi Aramco nowadays than just its American legacy, yet this is to be explored in the next section. Closing a cycle, the dissolution of Petromin was announced in a brief note in the local paper *Arab News* only after the death of King Fahd in fall 2005 (Hertog 2008, 645).

Throughout time, *international oil prices* have not markedly shaped the context. While in the 1970s soaring oil prices might have incentivized host governments to increase participation in the upstream sector (Tétreault 2008), upward and downward trends in oil prices ever since seem to have had no real impact on the upstream sector policy pursued by Saudi Arabia. Instead, Saudi Arabia has been able to shape the global oil price as it has deemed suitable. As the world's "swing producer" with 85 percent of OPEC's spare capacity, the Saudi state has had "a disproportionate influence over global oil prices" (Bronson 2006, 22). In this position, "since the mid-1970s, Saudi Arabia has ensured the free flow of oil at reasonable prices" (ibid., 3), which has informed and substantiated not just its standing in the international energy market but also its relations with major global powers.[9]

All 22 interviewees concurred about the leverage that Saudi Arabia has within OPEC. Three of the interviewees (think-tank, November 29, 2011; consulting, January 23, 2012; academia, April 12, 2012) pointed out that, especially in the past decade, Saudi Arabia has actively tried to maintain oil prices within an acceptable price band and balance out production in OPEC. Higher oil prices incentivize the development of alternative forms of energy, which are thus likely to reduce the demand for oil in the future. Therefore, opportunity costs for the development of alternative energies need to remain high lest gloomy prospects might loom large for an oil rentier state, which seeks to produce at the same rate (if not higher) for the next 50 years (government, January 24, 2012).

In conclusion to the *contextual factors*, the favorable geology and the high technical know-how of the NOC build indisputable strengths for the Saudi oil sector. The latter is a consequence of Saudi Aramco's American legacy but also of the political economy of the Saudi oil sector, which is to be discussed more narrowly in the next section. Neither a threat (not even during the bust period of the 1980s and 1990s) nor an opportunity, the global oil price in the past decade has in fact been steadily influenced by Saudi Arabia.

5.3.2.2 Domestic constraints

The attention is now moving to the main domestic constraints of political and economic nature, corresponding to the analytical framework presented in Chapter 3. These are the extent of oil dependence and the degree of executive constraints. Both of them are addressed systematically below.

Based on World Bank data, the *oil dependence* of Saudi Arabia over the past two decades has been consistently high and rising since the early 2000s (over 40 percent). Known as a redistributive state, the state's legitimacy "has been built around its capacity to distribute rent to different segments of society" (Marcel 2006, 107), as discussed at length in section 5.3.1 above. This was reinforced by an interviewee from the academia (March 18, 2012) with regard to the present situation: "the ability of the ruling family to redistribute and to maintain that distribution of income to continue to buy support and to ensure that opposition doesn't develop is absolutely crucial and it is oil and gas which really facilitate that." Quite innovatively, a government representative used the word "bedoukratia"—that is, the rule by the Bedouins—to characterize the politico-economic system of Saudi Arabia.

Instead of a government living off taxes collected from the population (i.e. the "production state" in Luciani 1990), the King as the custodian of the national wealth distributes some share of the oil money to the population. This trickle-down economics renders accountability of the ruling elites futile. Only *ijma´* (consensus) and *shura* (consultation) bring some democratic flavor to an otherwise arbitrary decision-making process (Hourani 2010). In short, "it is tribal politics extrapolated to the modern national stage" (government, February 14, 2012), where allocation and cooptation often pave the way to consensus. This has been confirmed by two other government representatives (February 1 and 21, 2012) who have had hands-on experience with the Saudi system.

Based on the typology proposed by the PFC Energy (2011, 48) and presented in section 2.2.3.1, Saudi Arabia is an "allocation state" primarily interested in "the redistribution of revenues for economic development, subsidies and government services." The institutions in the petroleum sector have to deliver correspondingly in order to match the state objectives. Formally, *the political economy of the Saudi oil sector* follows the Norwegian model of administrative design (Thurber *et al.* 2011), which includes a policymaking body (The Ministry of Petroleum and Mineral Resources), an oversight body (The Supreme Council for Petroleum and Mineral Affairs (SCPM) and an executive body (Saudi Aramco). Though a tripartite institutional design as endorsed by the World Bank in reliance on the Norwegian model, the Saudi oil sector is actually run by the Ministry of Petroleum and Mineral Resources and Saudi Aramco (industry, December 8, 2012). One of the interviewees with substantial experience both in consulting and industry turned out to be particularly sensitive about this one-size-fits-all template, which fails to take note of the power structures on the ground and leads to artificial institutions such as the Supreme Council for Petroleum and Mineral Affairs.

The Ministry sets the policy in the sense of establishing a price band, the production capacity and spare capacity. This way the "swing producer" status can be sustained. The policy follows the broad strategic guidelines provided by the King in consultation with advisers and other members of the ruling family (academia, March 7, 2012). Stevens confirms this information in his recent case study on Saudi Arabia while showing that "[t]he ruling family has ultimate control in the oil sector, but as a practical matter it delegates authority to a well-developed system of public administration" (2012, 187). In other words, the Ministry is the shareholder's representative both with Saudi Aramco and at the OPEC meetings. One of the academics (October 28, 2011) depicted the Saudi state system as binary in the sense that "everything that is on the security and defense side is controlled by the royal family, while anything that is related to oil, industrialization, chemicals, finance is de facto controlled by very prominent and efficient civil servants; that is, by commoners and only exceptionally by royal members." He concludes that "[o]ne of the major qualities of the ruling family is that it knows what it doesn't know. That means that they know well that if they get involved in oil they'll screw it up. So they let the specialists run the show."

Consequently, in the oil sector the Ministry and Saudi Aramco are the institutions running the show. In short, the Ministry disposes and Saudi Aramco proposes. The company is given full autonomy in developing and implementing the systems on the ground. Among all the 22 interviewees, it turned out to be indisputable that Saudi Aramco is a company managed on merit, insulated from political interference and enjoying an almost unique level of operational autonomy among NOCs globally.

In her comparative work of five Middle Eastern NOCs, Marcel depicts Saudi Aramco as a "fortress culture," "a national oil company operating in isolation from the rest of Saudi society" (2006, 63). Three of the interviewees (academia, October 28, 2011 and February 9, 2012; industry, December 8, 2012) further emphasized that there is not a single royal member in Saudi Aramco, not even on the Board of Directors (BoD). In this respect, the interview at Saudi Aramco shed light on the structure of the BoD, which is chaired by the Minister of Petroleum and Mineral Resources Ali Al-Naimi and contains three categories of individuals. Notably, all the

Saudi appointees on Saudi Aramco's BoD are from the public sector. It is difficult to say what their saying is and also, to what extent their influence might be compensated by the international external (non-Saudi) board members. For names and details, please see Table 5.1 below.

Corresponding to the Western model, Saudi Aramco management reports to the BoD for approval of the five-year business plan, capital programs etc. Final endorsement needs to be given by the SCPM (Saudi Aramco, December 13, 2011), yet this last stage is purely formal (in reference to 14 of the 22 interviewees).

For the approval of the company's plans and programmes, the BoD observes not just commercial criteria but instead, it takes a more comprehensive approach known as the "golden quadrant." This was presented by Saudi Aramco's representatives (December 13, 2012) and is also addressed by Stevens in several of his works (2008 and 2012). More explicitly, each project of Saudi Aramco is evaluated along two core dimensions: commercial prospects (in the sense of revenues generation and secure supply of global markets with petroleum) and social benefits (the pursuit of non-commercial objectives such as forward and backward linkages for the Saudi economy and society).

The contribution of Saudi Aramco to non-oil development is not as large and diverse as it used to be in the beginnings of the modern Saudi state when "there were no other institutions to execute and the options were limited to Aramco" (think-tank, November 29, 2011), whether hospitals, streets or

Table 5.1 The Board of Directors at Saudi Aramco (August 25, 2012)

4 members of corporate management	• The President and CEO, Khalid A. Al-Falih; • Senior Vice-President of Industrial Relations, Abdulaziz F. Al-Khayyal; • Senior Vice-President of Engineering and Project Management, Salim S. Al-Aydh; • Senior Vice-President for Upstream Operations, Amin Nasser.
4 Saudi nationals	• The Finance Minister, Ibrahim Al-Assaf; • The President of the King Abdul Aziz City for Science and Technology (KAUST), Mohammed I. Al-Suwaiyel; • The Chairman of the Capital Markets Authority, Abdul Rahman A. Al-Tuwaijri; • The Rector and Chief Executive of the King Fahd University of Petroleum and Minerals, Khaled S. Al-Sultan.
3 non-Saudi directors	• The former Chairman and Chief Executive of Chevron, David J. O'Reilly; • The former Chairman of Royal Dutch Shell, Sir Mark Moody-Stuart; • The former Executive Vice-President of the International Finance Corporation and Managing Director of the World Bank, Peter Woicke.

Sources: Saudi Gazette, August 25, 2010 and Saudi Aramco's official website

schools had to be built, bridges to be mended, Saudis to be hired or trained etc.—information confirmed by 15 other interviewees from all six different fields. Currently, the corporate social responsibility (CSR) activities of Saudi Aramco are still numerous yet more cohesive in their character. They tend to focus on research, business and academic training (for example, the creation of KAUST, Wa'ed Entrepreneur Development Initiative or the Dhahran Techno-Valley initiated by King Fahd University of Petroleum and Minerals), with other activities being marginal (Saudi Aramco's official website).

The Western pattern of governance is further reinforced by Saudi Aramco's budgeting scheme. The distribution of oil revenues between the NOC (with its capital and operational expenditures) and government is of high relevance to the sector's performance. In this regard, the World Bank shows that Saudi Aramco follows "the corporatized model" (Audinet *et al.* 2007, 16f.), as also revealed in section 2.1.1. Apart from the interviewees at Saudi Aramco (December 13, 2011), only two others (academia, October 28, 2011 and industry, December 8, 2012) were familiar with the budget scheme of Saudi Aramco and/or confident to answer such questions. Above the necessary budget to sustain the operations and develop the necessary systems, 80 to 85 percent of the profits are transferred to the Ministry of Finance as royalty and taxes while the remaining amount is either retained by the company to fund the capital programme or also goes to the government as dividends. Financially, Saudi Aramco is and has always been in the position of being treated like an IOC.

Overall, the company is well equipped both technologically and financially to run the operations by itself on the ground. Due to its both financial and non-financial contribution to the Kingdom, Saudi Aramco has had a high political standing and when needed, political teeth.[10] "The quintessential Aramco-reared oil functionary since the tender age of 11" (academia, February 9, 2012), Al-Naimi has enjoyed both King Fahd's and King Abdullah's complete trust, and has been the Minister of Petroleum and Mineral Resources since August 1995. His deep knowledge of Saudi Aramco, personality, negotiation and policymaking skills have made him, for eight of the interviewed experts, the lynchpin of the Saudi oil sector.

In this context, there seems to be no actual room left for the SCPM. Created in 2000, the SCPM's formal mission was to endorse the five-year operating and investment plans of Saudi Aramco, and to develop and approve the long-term strategy for the petroleum and mineral industry. The SCPM has had a number of royal members on its board, but their views have not necessarily been aligned with the King's thinking (Myers Jaffe and Elass 2007, 52). In his examination of the Saudi oil system, Stevens also confirms that "for most of the time the SCPM simply approved whatever was presented by the board of Saudi Aramco. It has never acted to stop any program of the company or any of its activities" (2012, 187).

The reaction of eight out of the 22 interviewees was evasive as to the SCPM's actual mandate, yet all the rest strongly agreed that the SCPM has never taken a proactive role. In short, "[t]he Supreme Council for Petroleum and Mineral Affairs is a non-entity" (academia, October 28, 2011).

To wrap up, "Saudi Aramco is the goose that lays the golden egg" (government, February 14, 2012) in a system which is highly dependent on oil revenues. Keeping Saudi Aramco walled-off has helped the company focus on the operational side of the business and deliver on it successfully. It has proven to have the probity to manage the operations in such a way that it has kept up the confidence of the government and maintained the monopoly over the oil upstream sector for decades. In the absence of any executive constraints, the Ministry of Petroleum and Mineral Resources has assumed the policymaking and partly, oversight role in a professional manner. For all that matter, the oil upstream sector in Saudi Arabia is run efficiently.

5.4 Open upstream sector: Abu Dhabi

This second case study focuses on Abu Dhabi among the seven emirates in the UAE because this is the genuine oil producing emirate. While Abu Dhabi owns 94 percent of the total oil reserves in the UAE, Dubai disposes of only 4 percent, with the remaining 2 percent being shared between Sharjah, Ra's al-Khaimah and Fujairah. The other two emirates, namely Ajman and Umm al-Quwain, have absolutely no commercially proven reserves (EIA 2013b, 3). Abu Dhabi is unique across the emirates not just because of the size of its resource endowment, but also due to its economic development path and projection of power in the UAE. The history of modern Abu Dhabi is the history of an oil-rich emirate which pushed for the creation of the UAE federation in 1971 and has retained the leadership ever since. For a better understanding of the institutional structures present in the oil sector of Abu Dhabi nowadays, a short historical account is essential. This draws on secondary sources and is followed by the discussion of the analytical framework in the context of Abu Dhabi based on primary data and when existent, also secondary sources.

5.4.1 Historical contingencies

The Al-Nahyan family has ruled over Abu Dhabi since its very beginnings— that is, since the discovery of water on the island in about 1761 (Davidson 2009a, 6).[11] For over 200 years now, the Al-Nahyan have been the ruling section of the Bani Yas tribe—"traditionally a heavily beduin, nomadic tribe which has occupied some of the harshest desert terrain of the Arabian peninsula" (van der Meulen 1997, 21) and which, as of the 1990s, comprised 27 different sections or families.[12]

The first signs of consolidation of the sheikhdom[13] came only in the second half of the nineteenth century with the reign of Sheikh Zayed bin Khalifa al-Nahyan, also known as Zayed the First (1856–1909). The late 1890s strengthened the influence of Britain in the lower Persian Gulf, which was less interested in colonizing these territories but more in maintaining the maritime supremacy of British East India Company over the trade routes between Bombay and Basra (Davidson 2005, 51ff. and 2009a, 11; Kechichian 2008, 280f.). To prevent the piracy raids undertaken mostly by the Qawasim as well as the rise of the Ottoman and Persian sway in the lower Gulf, the British entered into several "trucial agreements" with the Arab sheikhdoms which gave them the status of British protectorates (van der Meulen 1997, 50).

These truces inspired the name of "Trucial Sheikhdoms" or "Trucial Coast" for the Arab sheikhdoms in the lower Gulf (Kechichian 2008, 280). While the first rounds of truces dating from 1835 and 1853 were lax and envisaged maritime peace with the promise on the British side to watch for the external security of the trucial sheikhdoms, yet without interfering in their domestic affairs, the treaty of 1892 brought about economic isolation in the sheikhdoms and heavily impacted the early stages of the petroleum industry in Abu Dhabi. More explicitly, the trucial sheikhs were stripped of their powers to enter agreements with non-British parties and to sell or mortgage any domestic territory without the approval of the British Political Resident from Sharjah (Cordesman 2007). This would later compel the leader of Abu Dhabi to grant the first oil concessions to the main British oil company at the time.

Before the discovery of oil, Abu Dhabi had but a subsistence economy, based on date cultivation, fishing, nomadic animal husbandry (especially camels), traditional manufacturing (in particular, rugs, carpets, gowns, tents, daggers and swords) and pearling (Al Sadik 2001, 208; Butt 2001, 231f.; Davidson 2009a; Shihab 2001, 249). The growth of Abu Dhabi from an extremely poor desert sheikhdom—still as late as the mid-twentieth century—to the second-largest economy in the Gulf Cooperation Council only five decades afterwards (Koch 2011, 174) is mind-boggling. It is no doubt that oil changed the life of Abu Dhabians completely—yet not from the very beginning.

The first oil concession was signed in January 1939 for a period of 75 years by Sheikh Shakhbut bin Sultan Al-Nahyan with Trucial Coast Petroleum Development Company (PDTC),[14] a subsidiary of the Iraqi Petroleum Company (IPC), which was 51 percent owned by the British government (Al Fahim 1995, 43; Davidson 2009a, 31f.; Suleiman 1988 and 2007b, 775). The concession covered the whole onshore and offshore areas of Abu Dhabi and followed the provisions of the truce from 1892, which interdicted Abu Dhabi to close agreements with non-British parties. Sheikh Shakhbut received a signing fee of 300,000 rupees and the promise of a fixed annual income of 115,000 rupees. While exploration was postponed due to the commencement of World War II, the first geological surveys made PDTC retain all the onshore areas and relinquish the offshore in the early 1950s (Al Fahim 1995, 43; Davidson 2009a, 32). A second concession was granted in 1953 to D'Arcy Exploration Company[15] to set up the Abu Dhabi Marine Areas Company (ADMA) in charge of all the offshore areas for a period of 65 years (Suleiman 2007b, 775).

The first oil shipment left Das Island for Europe on July 5, 1962 from the offshore Umm Shaif field (discovered in 1959). From the onshore, exports started in December 1963 when the first tanker from the oilfield at Bab (discovered in 1953) left from Jebel Dhanna (Al Fahim 1995, 120; Rai and Victor 2012, 482f.). Only in the late 1960s/early 1970s did oil production in Abu Dhabi reach considerable levels, at about 700,000 barrels/day (Suleiman 2007a, 25).

Despite the oil revenues flowing into the sheikhdom, Sheikh Shakhbut did not invest anything in new infrastructure nor domestic economic development, while foreign operators did not care either. Consequently, the life of the Abu Dhabians did not improve at all during the 1950s and early 1960.[16] In his insightful account of Abu Dhabi's evolution "From Rags to Riches," Al Fahim wrote:

> We lived in the 18th century while the rest of the world, even the rest of our neighbours, had advanced into the 20th. We had nothing to offer visitors, we had nothing to export, and we had no importance to the outside world whatsoever. Poverty, illiteracy, poor health, a high rate of mortality all plagued us well into the 1960s.
>
> (1995, 88)

As basic physical infrastructure, financial organization and human capital were fundamentally missing in Abu Dhabi, oil companies had not just to bring in foreign personnel, who would have a long-lasting effect on the structure of the labor force to this day, but also to source their various supplies from Dubai. The Abu Dhabians were witnessing a revolution in their own sheikhdom without partaking of the benefits.[17] The primitiveness of Abu Dhabi in the twentieth century is striking when acknowledging the fact that not only had they nothing like electricity, running water, paved streets or concrete,[18] but the first school was opened in 1958 (Davidson 2009a, 35) and the first hospital came under construction in 1962, to be opened only in 1967 (Al Fahim 1995, 64).

Times changed with the succession of the Sheikh Zayed bin Sultan Al-Nahyan or Sheikh Zayed the Second, also known as the "Father of the Nation" (Koch 2011). The title is telling given that through distribution of the oil wealth and abrogation of all taxes, the new ruler "went far further than the tribal subsidies of the past and all but eliminated poverty in their sheikhdom" (Davidson 2009a, 44). He further kept his promise to the Al-Nahyan family members, who supported him in removing

Sheikh Shakhbut, and on getting a grip over power he appointed them to the most important state offices. In 1966 Sheikh Zayed set the basis for the new modern administration of Abu Dhabi as he appointed his brother Khalid, and the latter's son, as well as six of his cousin Muhammad bin Khalifa's sons, to important state posts (Herb 1999, 137).[19]

The Al-Nahyan thus managed to keep and strengthen their rule over the sheikhdom as the new patriarch Zayed co-opted not just the most powerful family members, but also the main sections of the Bani Yas tribe and other key tribes into the decision-making process of modern Abu Dhabi. "It is in noting the tribal affiliations of the nationals occupying these positions that a clear picture of the role of tribal and kinship ties in Abu Dhabi politics emerges" (van der Meulen 1997, 89). The tribes on which the Al-Nahyan have relied throughout time, particularly because of the loyalty shown in intertribal conflicts, have been the Manasir, with whom Bani Yas also shared their original home region in the Liwa Oases, the Awamir and the Dhawahir. A fourth tribe, the Za'ab, builds "a very recent addition to the base of support for Abu Dhabi's Bani Yas" (ibid., 91) since 1968 when Sheikh Zayed invited them to join the population of Abu Dhabi.[20]

The announcement of Britain's complete withdrawal from the region by 1971 brought the Trucial States to a crucial moment. Due to the rudimentary governments and poor defense capabilities, the security of the sheikhdoms of the lower Gulf looked dire. The need to come together and ensure some degree of security became clear.[21] A visionary mind, Sheikh Zayed, who was already contributing 80 percent of the total budget to the Trucial States Development Fund, pushed for the creation of the federal state. After long struggles between the ruling families of the lower Gulf emirates, an interim British-drafted constitution was finally signed and the United Arab Emirates (UAE) was created on December 2, 1971.[22] Given the economic leverage enjoyed by Abu Dhabi, Sheikh Zayed took over the presidency of the federation.

Drawing on the CIA World Factbook 2012, the UAE currently has one of the highest GDP per capita worldwide: $47,700. Yet, "significantly, if one were to pursue the GDP per capita line of research on an emirate-by-emirate basis, in Abu Dhabi it could be over $275,000 (or $17 million in reserves per national)" (Davidson 2007, 37). Also, the skyline of Abu Dhabi nowadays is a proof of its economic preeminence, which is stupendous given its misery from not long ago. Abu Dhabi invests a lot not just domestically in oil-related industrialization (e.g. metals, fertilizers, petrochemicals etc.),[23] high-tech heavy industries and future energy industries, but also overseas through the Sovereign Wealth Funds (SWFs).[24]

Notably, Abu Dhabi's population profits quite a lot from the oil wealth since the Al-Nahyan ruling family is known as an allocative institution. Distributions to Abu Dhabians take three main forms: direct wealth transfers and provision of free services, low-cost or free housing, and substantial employment and business benefits (Davidson 2009a, 128). Yet not all the

Emiratis are so well-off and the Arab Spring has showed some cracks in the surface. On the one hand, several relatively poorer citizens from the northern emirates have tried to voice their dissatisfaction with the current economic wedge between Abu Dhabi, Dubai and the rest of the emirates. On the other hand, some highly educated citizens from the national population, which make up less than 20 percent of the total UAE population, have manifestly asked for threefold freedom—of speech, of the media and political freedom[25]—and have sought to make the case against corruption, lack of transparency and human rights abuses. Such demands have resulted in tens of arrests and political imprisonments until this point in time.

Similarly to the Saudi-style spending spree, the UAE has tried to buy off the population under different forms. More specifically, the Al-Nahyan have raised the salaries in the public sector by up to 100 percent, increased welfare benefits by up to 20 percent and dispensed $2.7 billion for poorer nationals with outstanding loans (Davidson 2012). Politically, they expanded the electorate for the Federal National Council (FNC)[26] elections in September 2011 following the petitions early that year for a fully elected parliament and a true constitutional monarchy. More worryingly, in May 2011, the Crown Prince of Abu Dhabi employed a foreign private military company to create a secret 800-member private army of Columbians and South Africans to be deployed in case of domestic protests, internal unrest and/or infrastructure attacks (Mazzetti and Hager in *New York Times*, May 14, 2011). Whether it was the largesse of the Al-Nahyan family, the people's fear of mercenaries or the slight political opening, it is not clear how the UAE has managed to circumvent the Arab Spring, and also whether the future is going to be as politically and economically stable as the recent past.

5.4.2 The analytical framework against Abu Dhabi's empirics

Drawing on interviews with 16 experts, this section presents the empirical facts about the oil industry of Abu Dhabi and its upstream sector policy. The discussion below follows the structure of the proposed analytical framework. First, the contextual information about the geological conditions and technological know-how in the emirate of Abu Dhabi is presented and the role of international oil prices for the upstream control structures is investigated. Second, the economic and political elements which may domestically constrain the ruling elites in their choice of upstream sector policy are addressed.

5.4.2.1 The context

The UAE holds the seventh-largest proven oil reserves worldwide, of about 97.8 billion barrels. Of these, almost 92 billion barrels are located in Abu Dhabi (EIA 2013b). In 2011, the UAE (read: Abu Dhabi) ranked seventh among the world's largest crude oil producing countries, at a production rate of 2.69 million barrels per day. The goals of the Abu Dhabi National Oil Company (ADNOC) are to increase oil production capacity to 3.5 million barrels per day by 2017 (ibid.). This acceleration of oil production is quite surprising for Abu Dhabi given its typically low depletion rate (IO, January 24, 2012)[27] and the worsening *geological conditions*, as pointed out by six of the 16 interviewees (academia, March 7, 2012; think-tank, February 22, 2012; consulting, March 16, 2012; government, February 14 and 21, 2012; industry, December 8, 2011).

In 2005, the ranking of oil producing countries by upstream costs placed Abu Dhabi under the medium-cost oil producers, with $4.80 per barrel in total, out of which $3 go for exploration and $1.80 for production (Al-Attar and Alomair 2005, 250). Yet this is currently drastically changing. Until recently, "exploration risks have been very low (. . .). Driven by depletion of the country's 'easy' fields and also the political desire to boost output (and earnings), a new era of complexity and risk has arrived, confronting ADNOC with the need to boost investment in technology and expertise" (Rai and Victor 2012, 480). Given the limited prospects of major oil discoveries, the capacity expansion is to be based on enhanced oil recovery (EOR) techniques from the existing oilfields in Abu Dhabi (EIA 2013b, 5).

The centerpiece of Abu Dhabi's offshore oil industry is the Zakum system, which produced 30 percent of the country's total output in 2010. It is the third largest oil system in the region and the fourth largest worldwide. It includes the Upper Zakum and Lower Zakum oilfields. The former currently produces approximately 550,000 barrels per day, with recoverable reserves of 16 to 20 billion barrels, and it is operated by one of the offshore companies—Zakum Development Company (ZADCO). Created to develop Upper Zakum, ZADCO is currently owned by ADNOC (60%), ExxonMobil (28%) and Japan Offshore Development (Jodco) (12%). Due to the complex structure of the reservoir, the expansion plans by 200,000 barrels per day from Upper Zakum are estimated to cost $10–12 billion and involve advanced technical expertise for directional drilling. This concession was due to expire in March 2026 (PFC Energy 2011) but has been extended to 2041.

In turn, Lower Zakum oilfield, with a production capacity of 300,000 barrels per day, is run by the other offshore company, Abu Dhabi Marine Operating Company (ADMA-OPCO). This is owned by ADNOC (60.04%), BP (14.66%), Total (13.3%) and Jodco (12%) and it operates other offshore fields like Umm Shaif, Umm al-Dalkh, and Nasr/Umm al Lulu. The then production capacity of ADMA-OPCO, at approximately 560,000 barrels per day, was to be increased by 165,000 barrels per day by 2013. This concession expires in 2018 (EIA 2013b; PFC Energy 2011).

Finally, the onshore operations led by Abu Dhabi Company for Onshore Oil Operations (ADCO), i.e. the third major concession in Abu Dhabi, account for 55 percent of the country's total production. ADCO is owned by ADNOC (60%), ExxonMobil (9.5%), Total (9.5%), Shell (9.5%),

BP (9.5%) and Partex (2%). The major onshore oilfields under this concession include the Bu Hasa (600,000 barrels per day), Bab (320,000 barrels per day) and the Sahil, Asab, and Shah (SAS) fields (385,000 barrels per day). ADCO's production capacity should increase by 200,000 barrels per day in 2014. ADCO is Abu Dhabi's main onshore producer which— upon expiration of the concession in January 2014—continued the production activities solely on behalf of ADNOC (i.e., without international companies). Negotiations with IOCs from the U.S., Europe and Asia are currently ongoing, and the bidding process may last until January 2015 (DiPaola 2014). Its outcome is momentous for the future of the emirate's oil industry.

Notably, these three major concessions are the successors of the initial concessions signed by Sheikh Shakhbut in 1939 with the PDTC, and in 1953 with ADMA (see section 5.4.1). The former changed the name to Abu Dhabi Petroleum Company (ADPC) in 1962 and then to Abu Dhabi Company for Onshore Oil Operations (ADCO) in 1979. On January 1, 1973, the government of Abu Dhabi acquired 25 percent equity in the company and increased it to 60 percent one year later. The interest was vested in the hands of the NOC, ADNOC, which was created from null through Law No. 7 of November 27, 1971, at a time when the trend of state participation in the region was on the rise as the governments in the producer countries were trying to take control over their national resources (Suleiman 2007b, 775ff.). Apart from the shares in the onshore company, in 1973 the government of Abu Dhabi also presented ADNOC with 25 percent interest in ADMA and raised it to 60 percent one year later. In 1977, ADMA-OPCO was created as a second subsidiary of ADNOC to succeed ADMA as an operator of the concession (ADMA-OPCO website and Suleiman 2007a, 67). The same year, the relinquishment of the Upper Zakum oilfield by ADMA-OPCO led to the creation of the third major joint venture, ZADCO. Additionally, several smaller concessions were awarded to other operators in two stages: 1967 to 1971 and 1980 to 1981.[28]

Notably, the changing geological conditions coupled with the capacity expansion plans have elevated the technological requirements to produce the fields in Abu Dhabi. Concerning the level of *technical and technological capabilities*, ADNOC has depended on IOCs and service operators throughout its entire existence and ironically, this reliance has increased over time as the reservoirs are growing mature and operations are becoming more complex. By comparison with other NOCs which had had time to gain experience until the 1970s, ADNOC was in its very early stages and incapable of operating independently (academia, November 18 and 30, 2011, February 10, 2012, and March 7, 2012; government, February 14 and 21, 2012). This explains why among all the countries in the region, Abu Dhabi uniquely retained foreign partners in its upstream sector—albeit to a limited extent, i.e. 40 percent equity in the concessions—despite the wave of full-blown nationalizations.

It is impossible to recount exactly how the technical and technological capabilities of ADNOC have evolved over time. All 16 interviewees emphasized that by contrast to Saudi Aramco, which has become increasingly transparent over the past 10 years and has disclosed a lot of data on its operations, the broad characteristics of the oilfields and its technical know-how, ADNOC has remained utterly secretive about all these aspects. While the company reports on its website to have used cutting-edge seismic surveys and geophysical studies in its operations, it is not clear to what extent these tasks have been performed by ADNOC itself or instead by foreign operators.

Opinions vary a lot. Out of the 16 interviewees, only seven of them were willing and/or felt knowledgeable enough to address the question about ADNOC's technical capacity. Their responses cover the full spectrum. More specifically, two of the interviewees (academia, March 7, 2012; think-tank, February 22, 2012) emphasized that ADNOC is operationally not very active but it merely oversees the IOCs. To quote one of these experts (academia, March 7, 2012), "ADNOC continues to be a company that is not doing much. It's non-existent. They don't do anything apart from watching their joint venture partners." In turn, three other interviewees (academia, November 30, 2011 and February 10, 2012; consulting, March 16, 2012) alleged that ADNOC has been operating some minor fields on its own, yet it has remained dependent on foreign expertise for the main operations. The stance of these three interviewees is well captured by the consultant's response (March 16, 2012): "It is fair to say that ADNOC's technical capability is fairly weak—it is not completely incapable but it is certainly not up to the standard of Saudi Aramco in terms of operating on its own." Eventually, two other experts (industry, December 8, 2011 and academia, March 18, 2012) depicted ADNOC in a more positive light as a relatively efficient and capable operator lately. The industry professional cautioned that "its technical prowess and ability to manage the IOCs should not be underestimated. ADNOC is no longer a façade." While the experts do not seem to fully concur with each other, it should be noted that the majority of the interviewees have not refuted ADNOC's operational capacity and technical competencies completely. Along the same lines, Rai and Victor emphasized that ADNOC might have failed in its R&D activities, but has proven itself in acquiring and utilizing cutting-edge technology with the assistance of its foreign partners and other oil service companies (2012, 492ff.).

Throughout the past four decades, *international oil prices* have had no actual impact on the control structures in the oil upstream sector of Abu Dhabi. Since 1974 the proportion of state versus foreign private control in Abu Dhabi's oil upstream has remained the same (60:40), irrespective of the oil price fluctuations. In OPEC, the UAE (read: Abu Dhabi) has followed the Saudi way and has behaved like a "price dove" (confirmed by 14 of the 16 interviewees, with two abstaining). In other words, it has not been incentivized to militate for higher prices, but quite the contrary—and that is in order to preserve their oil markets in the long run.

In conclusion to the *contextual factors*, the recent history of modern statehood in the UAE and the newly created NOC ADNOC as late as 1971 have restricted the options of Abu Dhabi's leaders in the design of the oil industry. The support of foreign oil companies has been imperative for the oil industry in Abu Dhabi to grow and get established. Given the increasing complexity of the geological conditions in Abu Dhabi as well as the moderate technical expertise acquired by ADNOC in time, foreign assistance continues to be required for the maintenance and expansion of oil production capacity. Consequently, the domestic geological and technological premises lie somewhere in the middle on the strength–weakness continuum introduced in the analytical framework (section 3.2.1). As for the impact of oil prices, there is no actual trace thereof during the past four decades.

5.4.2.2 Domestic constraints

The domestic constraints of economic and political nature are addressed next. Economically, Abu Dhabi has been a heavily oil-dependent emirate. Notably, the World Bank and IMF statistics on oil rents or the value of oil exports by country depict the economy of the UAE. At an aggregate level, since the late 1980s until the present, oil revenues for the UAE have not reached the threshold of 40 percent of GDP (Ross 2012). Given the fierce diversification taken up by Dubai as well as the insignificant oil proceeds to the overall GDP from the other emirates, the figures produced by the two abovementioned international organizations are not reflective of Abu Dhabi's degree of *oil reliance*. Despite various efforts of economic diversification,[29] the oil sector remains the powerhouse in the emirate of Abu Dhabi and tops well above 50–60 percent of the GDP, in reference to all 16 interviewees.

Following in the lines of history, Abu Dhabi is an allocation emirate where,

> the ability of the ruling family to redistribute income to buy support and ensure that opposition does not develop is absolutely crucial, and it is oil and gas which really facilitates this. It is oil and gas which makes the whole system of governance run. Any disruption to that would have immediate consequences for Abu Dhabi and the UAE.
>
> (academia, March 18, 2012)

The social pact between the Al-Nahyan and the Abu Dhabians has been corroborated by seven experts—from the academia (November 18 and 30, 2011, March 7, 2012), think-tank (February 22, 2012), consulting (January 26, 2012), government (February 14, 2012) and international organizations

(February 15, 2012)—as well as by scholarly work (Davidson 2009a and 2009b; Al Sadik 2001; Shihab 2001; van der Meulen 1997).

The institutions in the petroleum sector have to provide accordingly so as to sustain the ruling bargain of the Al-Nahyan in the Emirates. *The political economy of the oil sector* in Abu Dhabi has yet to be delimited from that of the UAE. Importantly, "[t]he individual emirates have jurisdiction over their petroleum affairs" (Suleiman 2007a, 60)—in other words, each emirate has sovereignty over its own resources and petroleum policy. The institutions at the emirate's level do not match the Norwegian model of administrative design (Thurber *et al.* 2011). Functions are relatively muddled in the sense that ADNOC is not just the executive body in Abu Dhabi but also oversees the operations of foreign oil companies in the joint ventures.

As regards policy-making and control/oversight of ADNOC in Abu Dhabi, these functions were vested in the Abu Dhabi Petroleum Department back in 1971 and taken over by the Supreme Petroleum Council (SPC) in 1988. At the federal level, there is a Ministry of Energy which has a very limited role. In theory, this should be the policymaking body at the UAE level, deciding over production quotas in the emirates, crude oil prices and representation in OPEC (Suleiman 2007a, 191 and Marcel 2006, 77). In practice, these decisions are made by the ruling leaders of Abu Dhabi—that is, the Al-Nahyan—given that for the past couple of decades this has been the actual producer emirate in the UAE (industry, December 8, 2011; think-tank, February 22, 2012; and consulting, March 16, 2012).

Since the late 1980s, SPC has been the supreme body of governance in Abu Dhabi's oil sector (information confirmed by ten interviewees, see also Rai and Victor 2012). It was created in an effort to streamline the petroleum sector by cutting down on bureaucracy and overcoming a conflict of personal interests between the chairman of the Abu Dhabi Petroleum Department, Mana Said Al Otaiba, and the chairman of ADNOC's BoD, Sheikh Tahnun bin Mohammed Al-Nahyan. The government of Abu Dhabi dissolved both units and supplanted them with SPC (think-tank, February 22, 2012 and academia, March 18, 2012). As an appointed body, SPC came to include the Ruler of Abu Dhabi as its chairman, the Crown Prince, a number of other ruling individuals and a few technocrats (consulting, March 16, 2012). In terms of functions, SPC has had to lay down "the Emirate's policy and its objectives in all sectors of the petroleum industry, in addition to issuing resolutions for implementing its policy, and follow up such resolutions until the achievement of the aspired results" (ADNOC website). To put this more precisely, it has set the strategy and goals for the petroleum sector and has further delegated the operational decisions to ADNOC. In Marcel's opinion, "the Abu Dhabi (and UAE) political system is highly centralized and cohesive. High strategy decisions are made by government (that is to say, the ruler), and these are executed by the NOC. Because of the unity of purpose, there is no need for government 'interference' in ADNOC's operations" (2006, 86).

The composition of SPC is highly relevant for the way in which the oil sector is run. The Emir's decree on June 25, 2011 amended the membership of SPC. Apart from the Ruler of Abu Dhabi, Sheikh Khalifa Bin Zayed Al Nahyan, who stayed on the chairmanship, currently there are:

- five members from the Al-Nayhan family (four brothers and one son of Sheikh Khalifa): Sheikh Sultan Bin Zayed Al-Nahyan, General Sheikh Mohammed Bin Zayed Al-Nahyan, Sheikh Mansour Bin Zayed Al-Nahyan, Sheikh Hamed Bin Zayed Al-Nahyan, and Sheikh Mohamed Bin Khalifa Bin Zayed Al-Nahyan;
- three members from the Al-Suwaidi merchant family: Mohamed Habroush Al-Suwaidi, Hamad Mohamed Al Hur Al-Suwaidi, and Abdulla Nasser Al-Suwaidi;
- one member from Al-Kindi merchant family: Mohamed Khalifa Al-Kindi;
- one member from the Al-Dhaheri family/the Dhawahir tribe: Jua'an Salem Al-Dhaheri.

(Source: ADNOC website)

The majority of members (six) in the SPC come from the Al-Nahyan ruling family while the other five positions are occupied by representatives of the two most relevant merchant families and respectively, one of the allied tribes. In reference to van der Meulen (1997)'s record of tribes and respective sections, Al-Suwaidi and Al-Kindi families belong to the same tribe like the Al-Nahyan, namely the Bani Yas tribe. Petroleum affairs are thus decided in the family and at most, "in the tribe." In other words, tribal affiliations remain extremely relevant for the allocation of political posts and economic patronage (see discussion in section 5.4.1). Commenting on the composition of SPC, one of the interviewees noted:

> the SPC is a way of bringing in a number of important and influential figures not just from the ruling family but from other major merchant families of Abu Dhabi as well (especially of the Al-Suwaidi and Al-Kindi families). It is a way of coopting and integrating potential rival influence into the heart of the regime. It is all about balancing between different interest groups and ensuring that a) one doesn't become too powerful but also b) that one doesn't feel left out.
>
> (academia, March 18, 2012)

This strategy has proven largely successful in annulling any form of political constraints in the emirate throughout time.

Rather interestingly, three of the academic experts cared to emphasize that the direct involvement of the Al-Nahyan in running the oil sector—through the SPC—and the role of individual personalities like Sheikh Zayed or Sheikh Khalifa, have actually been instrumental in shaping a stable political environment and, thereby, creating attractive conditions for foreign investors (academia, November 18, 2011, March 7, 2012 and April 12, 2012).

While all major policy decisions are made by the SPC and thereby, the ruling family, ADNOC is tasked with the operational side of the oil business, where it seeks to deliver on two main fronts. These are: first, the supply of Abu Dhabi's government with enough revenue streams to meet the emirate's and the UAE's economic development plans, and second, the coverage of the country's rapidly growing energy needs. ADNOC has constantly achieved both goals successfully, albeit with the support of IOCs, which makes it a reliable player in the institutional setup of Abu Dhabi's oil sector.

Notably, "the royal family is not really involved in the day-to-day management of the oil sector—that's done by loyal technocrats" (consulting, March 16, 2012). This is not to say that ADNOC is completely insulated from the cooptation and integration strategy pursued by the Al-Nahyan through the government apparatus. In the executive management of ADCO, ADMA-OPCO and ZADCO there are several members from influential families and tribes in Abu Dhabi (such as Al-Kindi, Alqubaisi, Al-Suwaidi, Al-Mazroui and others), as shown by the respective websites. Yet the employment of prominent figures in high positions should not necessarily be equated with lack of performance. ADNOC is the only NOC in the MENA region which penalizes poor performance among the employees (Marcel 2006, 59).

In fact, there are a number of paradoxes about ADNOC which might be the product of a mix of cultures: "a company with a strong commercial culture and modern management processes but slow, traditional and consultative decision-making procedures" (ibid.). Despite the strong hierarchy, each partner and level of management are consulted as part of the decision-making process, which becomes bureaucratic, lengthy and quite frustrating for the IOCs. Eventually, all significant decisions are still made at the top by a few key managers (academia, February 10, 2012 and consulting, March 16, 2012; see also Rai and Victor 2012, 491). Also, the company seems to be transparent in its accounting practices and open in sharing information with the shareholders (i.e. both foreign partners and the government of Abu Dhabi), but at the same time, extremely "wary of communicating with outsiders" (Marcel 2006, 59).

For each of the three concessions, there is a Joint Management Committee (JMC) which frames the operational targets, investment and development plans. Subsumed to the JMC, the BoD of each of the three ADNOC subsidiaries is charged with the oversight of the technical and financial performance. Both the JMC and BoD are composed of representatives of ADNOC and the foreign partners, yet "power in these institutions is concentrated in the hands of ADNOC's senior officials" (Rai and Victor 2012, 502).

Regarding finances, none of the interviewees was able to provide a clear answer about ADNOC's budgeting model. This might be also indicative of ADNOC's secretiveness. Drawing on a World Bank's report, ADNOC seems to follow the so-called "corporatized model," similarly to Saudi Aramco. More specifically, the revenues ensue to ADNOC's subsidiaries which first cover expenses and distribute the remaining profit under different forms such as royalty, taxation and dividends to the shareholders (Audinet *et al.* 2007, 16f.). As for the financial take of the foreign partners in the three main concessions, this "is determined essentially by the terms and conditions of the individually negotiated agreements concluded between government and IOCs" (Suleiman 2007a, 52f.).

By contrast with other NOCs, ADNOC has had a relatively confined national mission beyond its oil and gas operations (Tordo *et al.* 2011). In this respect, "ADNOC is much more focused on its core business than other NOCs although it does have a mission to pursue Emiratization—increasing jobs for nationals through training and hiring targets set by the SPC" (think-tank, February 22, 2012). The non-core activities are marginal and are primarily related to research and education through the Petroleum Institute and the ADNOC Technical Institute (industry, December 8, 2011; academia, February 10, 2012; think-tank, February 22, 2012; government, February 14 and 21, 2012; see also Rai and Victor 2012, 495).

All in all, given the overly powerful SPC, the relatively competent ADNOC and the constant technical support of the IOCs, it is not surprising that there is little room left for a Ministry of Energy at the UAE level. In short, "[t]he main function of the oil minister is to represent the Emirates at OPEC. (. . .) The ministerial post remains a 'political position'" (Marcel 2006, 104), wholly insignificant at the level of the oil-richest emirate.

Before drawing any conclusions, a few words are needed with respect to the uniqueness of the emirate of Abu Dhabi in the MENA region. Addressing the reasons for the absence of complete nationalization in the history of Abu Dhabi, interviewees provided two further major explanations.

First, eight of the 16 interviewees pointed to the young history of the country and its lack of technical capacity as compelling factors for letting the IOCs retain 40 percent equity in the sector and uphold the oil operations (think-tank, February 22, 2012; consulting, March 16, 2012; academia, November 18, 2011, November 30, 2011, February 10, 2012, and March 7, 2012; government, February 14 and 21, 2012). This compromise solution (60:40) was reinforced by the good relations of Abu Dhabi's government with the IOCs. Two small add-ons were brought to this historical explanation, as follows: the widespread sentiment (prevalent to this day) was that the oil industry was delivering on financial terms while the sector remained "under national control with foreign assistance" (consulting, March 16, 2012) and also, due to the small size of Abu Dhabi's population in the 1970s, there was no actual political pressure for nationalism and full nationalization (academia, March 18, 2012).

The second main explanation can be summarized as "diversification of security" and was suggested by three academic experts (November 30, 2011, March 18, 2012, and April 12, 2012). Tying foreign operators into the oil industry gives security guarantees to the host government as the IOCs and, most of the time, their countries of origin raise stakes in the local security and stability of the oil production in the respective producing country. Quite interestingly, one of the interviewees regarded this as a dimension of energy security for oil producers: "[t]hese countries like the UAE are trying to diversify their security because they realize they are in a volatile part of the world. For them, energy security is about maintaining the exports of oil and inflow of revenues to make the redistributive state continue" (academia, March 18, 2012). Along these lines, several scholars have underlined the vulnerability of the UAE and have pointed to various security risks such as the disputes with Iran about the occupied islands (i.e. Abu Musa, and the Greater and Lesser Tunbs) or a potential blockage of the key oil export routes like the Strait of Hormuz (Cordesman 2007 2008; Crystal 1998; Ulrichsen 2009a and 2009b).

Whether out of economic, socio-political and/or geopolitical reasons, the presence of foreign oil companies seems to have served the oil industry of Abu Dhabi quite well. In the absence of any executive constraints, the Al-Nahyan ruling family has kept a strong hold on the oil sector—in particular through the SPC since 1988 but also before, through the Department of Petroleum—and has passed clear operational decisions onto ADNOC. Having to build its internal and external capabilities from null, ADNOC has got enmeshed in bureaucratic and protracted decision-making processes and to this day, it has lagged in technological know-how and technical expertise. Yet these paucities have not impeded it from delivering on the major economic goals set by the government in a relatively successful manner by nourishing (fairly) good relations with the IOCs and reaping the technical benefits therefrom. For all these reasons, the oil upstream sector in Abu Dhabi is run in a way which has catered to the purposes of the Al-Nahyan and the wealth of the emirate's population reasonably well.

5.5 Conclusions to the case studies

The case studies on Saudi Arabia and Abu Dhabi teach us that there is a mix of explanations which account for the choice of the oil upstream policy. These explanations substantiate and complement (some of) the arguments outlined in the analytical framework.

Based on the two case studies, the role of both geological conditions and technical capabilities of the NOC can be validated. The extremely favorable geology in Saudi Arabia can be pitted against the medium and currently increasing upstream costs in Abu Dhabi to provide support for *Hypothesis 1*. Clear evidence has been further found for *Hypothesis 2*. More explicitly, the technical prowess of Saudi Aramco—following the smart nationalization of the American Aramco—stands the comparison to any IOC. By contrast, ADNOC is a company created from scratch in the early 1970s, at a time when the emirates were struggling for the creation of the UAE. Although technical expertise has been accumulated over time, ADNOC has remained dependent on the technical assistance of the IOCs given the increasingly complex geology. No role has yet been found for oil prices—with respect to *Hypothesis 3*—in shaping the oil upstream policy in either of the countries. Furthermore, both Saudi Arabia and Abu Dhabi have indisputably upheld high reliance on oil revenues over decades. Given that the extent of oil reliance is relatively comparable, it is surprising that Saudi Arabia has kept the oil upstream sector under full state control whereas Abu Dhabi has retained only 60 percent control thereof. On this note, the empirical evidence from the two cases is not necessarily confirmatory for *Hypothesis 4* as such.[30] The same is the case with respect to *Hypothesis 5* when the effect of low executive constraints is taken individually. Yet the case studies elucidate how oil reliance and executive constraints influence and reinforce each other in explaining oil upstream sector policies, as will be discussed below.

Beyond partial support to the relatively rough relations enunciated in the analytical framework, the comparative findings from the two case studies provide *four other potential explanations*. These are drawn from the wealth of insights rendered by the primary data, which was at times seconded by existing literature. As a matter of fact, apart from geology (exogenous factor) and the international oil price (without explanatory power in the case studies), the other three factors included in the analytical framework are strongly embedded in the historical pattern of development. These factors are not standalone. In this sense, the case studies help understand how the current status of technical capabilities, the constraints on the decision-making process as well as the structure of the national economy and thereby, the extent of oil reliance, are the product of different national historical paths.

First, the premises for the modern state in Abu Dhabi are widely different from those in Saudi Arabia. The former was a British protectorate and following several truces from 1835, 1853 and more importantly, 1892, it forfeited not only the decision over its foreign policy but also the power to close arrangements with non-British parties (or at least, without prior approval of the British Political Resident). This resulted in the forced concessions with PDTC and D'Arcy Exploration Company, both of which were at the time majority-owned by the British government. In turn, Saudi Arabia could select the operators of the concessions as it deemed suitable. IOCs had a largely different impact on the economic development of the Kingdom and respectively, the Emirate, before nationalizations. Aramco invested a lot in infrastructure and the build-up of the Saudi economy at a time when the modern Saudi state was taking shape. In turn, the oil industry in Abu Dhabi was initially kept separate from the indigenous population and foreign companies reaped off the benefits. The creation of modern Abu Dhabi started only in the late 1960s when the wave of nationalizations was already ongoing in the region and Sheikh Zayed understood to take the reins of power and domestic economic development in his own hands. In short, it can be postulated that the colonial history in Abu Dhabi created a serious delay in the economic development of the modern state and the beginnings of its oil industry.

Related to the latter aspect, a *second* explanation marking the different trajectory of the oil upstream sector policy in Saudi Arabia versus Abu Dhabi is the oil production start. While the first oil concessions were signed at around the same time in the two cases—that is, in the 1930s—in Abu Dhabi, production only started over 20 years later, in 1962. This leads to considerable impedance for Abu Dhabi not just in monetary and developmental terms, but also regarding the setback in technology and human capital for the oil industry. This further explains the different behavior of the two cases about ten years afterwards, when the issue of increased participation/nationalization was raised in the region: Saudi Arabia dispensed itself completely of foreign partners while Abu Dhabi could not. More explicitly, the indigenous capacity was lacking for Abu Dhabi to take over the petroleum industry completely and run the sector by itself. Also, the bargaining power of foreign companies in Abu Dhabi was still too high for them to stay on "retained royalty" terms like in Saudi Arabia. As a matter of fact, this question did not even come up but instead, 40 percent control rights were preserved.

Third, the interviewees pointed to the demographics as another potential explanation for the way in which the process of increased participation unfolded in the 1960s and 1970s. The small population of Abu Dhabi did not put political pressure for nationalization. As discussed in the historical section of the respective case study, the people were also not at that stage (size-wise, economically but also politically) to be able to raise resource nationalistic demands. By contrast, Saudi Arabia had a comparably larger population which was long capable of vying for their rights to the national resources (Vitalis 2006). Yet demographics may influence the policy choice in the oil sector through another mechanism. Demographics, together with territory, frame the power status of a country. Given the volatility of the region, small powers like Abu Dhabi are likely to opt for diversification of security by tying in foreign operators into their oil sector. As these companies and their countries of origin nourish strategic interests in the host oil-rich country, they are more likely to intervene in the case of military aggression by other states, civil conflict and other forms of national insecurity. In turn, given the power projected by Saudi Arabia in the Middle East region (and beyond), the respective government did not feel itself in a weak position so as to make concessions to foreign oil firms for the sake of national security.

Fourth, the two case studies point to the role of the ruling family as a strong explanation of the oil sector organization and its upstream policy.

Intentionally selected so that the limits on the executive are low and the degree of national oil reliance is high, Saudi Arabia and Abu Dhabi render the dynastic family instrumental in shaping the redistributive institutions and not least, the political economy of the oil sector. In reference to Dunning's (2005) inspiring work, the two cases exemplify that when executive constraints are low, the ruling elites may indeed purposefully sustain high oil reliance. Both in Saudi Arabia and Abu Dhabi, the ruling family took up and consolidated their hold on the main source of revenue in the early history of the modern state. By dispersing oil wealth, the ruling family destroyed existing elites, carved new social classes (in Saudi Arabia) or re-organized the importance of tribes and merchant families in the society (in Abu Dhabi) depending on the wealth distribution from the top. To this day, both the Al-Saud and the Al-Nahyan cultivate high oil reliance and uphold the system of patronage and subsidies. This is part of a well-thought-out strategy to secure their stay in power.

Based on the statistical analysis, it has been found that higher constraints on the ruling family and elites are likely to bring about more state control in the oil upstream sector. In the presence of low executive constraints, the case studies on Saudi Arabia and Abu Dhabi have further uncovered that first, the oil reliance may be nurtured and sustained at and from the top as a form of political rationality and second, that under this constellation of factors, the oil upstream policy choice may be more strongly conditioned by the contextual characteristics than otherwise. That is, when executive constraints are low and the oil revenues fund their stay in power, the ruling family (or more generally, the incumbents) is more prone to guide their upstream sector policy choice based on economic considerations. The domestic context, framed by geology and the technical capabilities of the NOC, is then expected to play a more substantial role in the decision over the upstream policy. Since the oil industry is the cash-cow which feeds into their strategy to stay in power (i.e. the oil reliance link), the ruling elites are likely to keep their options open as concerns the operations in the oil upstream sector (i.e. production only through the NOC or also through IOCs). This is why, in the absence of technical expertise to meet the geological requirements, ruling elites in oil producing countries are likely to bring in foreign operators and grant them control rights.

Both case studies further show that the ruling families have tailored the political economy of the oil sector quite shrewdly from an economic perspective—given that both oil sectors are run (fairly) efficiently. In Saudi Arabia, the family institution is not directly involved in the oil industry. The Al-Saud family has allowed the specialists—i.e. Saudi Aramco (with its American legacy) and the Ministry of Petroleum and Mineral Resources—to "run the show." The former is comparable to an IOC with respect to its R&D activities, level of expertise, budgeting scheme and operational autonomy. If there is a difference to a major foreign privately-owned oil firm, this is the strategic autonomy—that is, the lack of a strategic vision including, for example, operations or partnerships overseas. The Ministry is, in turn, the interface between Saudi Aramco and the ruling family, which frames the policies at the sectoral level as well as the dispositions for the NOC. In line with the idealistic Norwegian model of governance, there is a third pillar, the SCPM, in the political economy of the Saudi oil industry. Its board is mostly composed of royal members, yet the SCPM has no say in the petroleum policy or more generally, the running of the oil sector. This is again a confirmation of the limited involvement of the royal family. The King and his family have full confidence in the operational decisions and capacity of Saudi Aramco—which, if needed, is known to employ the services of foreign operators like Schlumberger or Halliburton under service contracts—as well as in the Ministry's bureaucratic delivery on its policy-making and oversight duties. Operationally fully controlled by Saudi Aramco and policy-wise run by the bureaucrats in the Ministry, the oil upstream business in Saudi Arabia is efficient.

By comparison, in Abu Dhabi/the UAE, the statehood was in its early stages when producer governments in the region decided to take control of their petroleum industries. Several IOCs were administering the oil resources of Abu Dhabi and the industry was still in its formative years. On these grounds, the Al-Nahyan family has not opted for the expropriation of the foreign oil companies' assets but instead, created an NOC from null to work with the foreign firms in all three concessions. Time proved the Al-Nahyan right in their 60:40 formula. This way, they ensured that they retained the final say in any decision concerning the oil industry while employing consultations with the foreign partners in the concessions. The sentiment in Abu Dhabi, shared both by the ruling elites and the people, has been that the foreign firms have been providing assistance in the oil sector while the control stays in the hands of the state. By contrast to Saudi Aramco, ADNOC is a mere executor with oversight of the concessions and thereby, foreign partners.

The supreme body of governance in the oil sector—arguably, more extensive in its scope than the Saudi Ministry of Petroleum and Mineral Resources—is the SPC. All strategic decisions in the oil industry (including those at the level of ADNOC-IOCs concessions) are taken by the Al-Nahyan, whose members dominate the SPC. In other words, in stark opposition to Saudi Arabia, the ruling family is directly and overtly involved in the running of the petroleum industry. This may have been a means in the 1970s to ameliorate for the lack of an established oil industry and capable indigenous people to "run the show," and at the same time, to keep oversight of the foreign partners with control rights in the concessions. The oil upstream sector policy has remained unchanged for over four decades now – the current phase in the onshore concession is transitory and IOCs are likely to partake in the operations by early next year – and the Al-Nahyan continue to be hands-on and the evident nucleus of power in the petroleum industry of Abu Dhabi.

To close the triangle prescribed by the ideal model of governance, it should be mentioned that the political economy of the petroleum sector in Abu Dhabi relies on the SPC and ADNOC plus IOCs. Although there is a third, (supposedly, policy-making) institution—the Ministry of Energy at the UAE level—this has no leverage on the oil policy at the emirate's level. All in all, with the Al-Nahyan at the apex of governance and ADNOC working with IOCs in the concessions, the oil sector in Abu Dhabi has performed well across time.

In conclusion, the case studies brought to light four alternative explanations of oil upstream policy choices which go beyond the ones comprised in the analytical framework and tested in the statistical analysis. These quantitative and qualitative findings need to be reviewed in the following chapter, before any amendments are made to the proposed analytical framework.

Notes

1 As will be discussed in the case study, in the absence of disaggregated data, it is legitimate to equate the production of the UAE to that of Abu Dhabi given that the latter is the actual oil producing emirate, with approximately 95 percent of all oil resources. The emirates of Dubai, Sharjah and Ras Al Khaimah have relatively insignificant reserves with oilfields running mature, while the other three emirates have no resources whatsoever. As Suleiman (2007a) puts this, "[s]ince Abu Dhabi is the largest oil producing emirate, with the most substantial oil reserves, the richest experience in the oil industry and the longest history of relationships with foreign oil companies, it is ideally suited to serve as the focus of a study of the development of petroleum resources in the UAE" (Back cover).

2 Population data is taken from CIA World Factbook, (status: September 18, 2012).

3 In the course of a Fulbright fellowship in Washington D.C. during 2011/2012 academic year, 33 experts were interviewed.

4 For an insightful discussion of this sampling technique, please refer to Biernacki and Waldorf (1981).

5 The quotes were cleared by two of the Saudi Aramco employees, who took part in the interview from December 2011. The permission to disclose the information was given in writing via email on April 11, 2013.

6 The official name dates back to September 22, 1932 (Long and Maisel 2010).

7 The founder of the royal house is Muhammad bin Saud bin Muqrin bin Markhan (1703/04–1765), who, aided by Muhammad bin Abd al-Wahhab, the leader of the Wahhabi reform movement, embarked on a campaign of political expansion under the banner of tawhid (i.e. strict monotheism) as of 1744 (Long and Maisel 2010, 28f.). From "the amir of Dir'iyya, a petty principality just down the Wadi Hanifa from Uyayna," the Al Saud family domain was expanded by the beginning of the nineteenth century, under Muhammad bin Saud's son and grandson, to cover most of the Arabian Peninsula (including Karbala in present southern Iraq, Makkah, Madinah and to the east down to Oman). Thus, "[i]n a relatively short period of about 60 years, the Saudi regime had been transformed from a tiny oasis principality to an important Middle Eastern state" (ibid., 29). The first state would only last for a short time because the Ottoman caliphate, deprived of considerable income due to the Al Saud's capture of the Muslim holy places, decided to invade Najd and destroyed Dir'iyya in 1818, thereby terminating the first Saudi state. The second Saudi state was rebuilt as of 1834 and it had grown

prosperous under the second rein of Faysal (1843–1865), Turki bin Abdullah's son. After his death, the infightings for the rule between the Al-Saud brothers weakened the state so that in 1871 the Ottomans reconquered al-Hasa in the east whereas Oman and the Trucial States in the south shook off the Saudi rule with the help of the British. For a complete history of the first and second Saudi state, refer to Long and Maisel (2010).

8 For a graphical representation of the Saudi government at its first meeting, see Figure 4 in Herb (1999, 92).

9 This is a topic which goes beyond the scope of this book. For this, please see Bronson (2006).

10 "The Saudi Gas Initiative makes for an interesting case of bureaucratic politics outside of the regular domestic political economy, in which turf-conscious Aramco technocrats managed to significantly undermine the government's strategy of opening" (Hertog 2010a, 139). The initial invitation launched by Crown Prince Abdullah to IOCs on September 26, 1998 to the residence of the Saudi Ambassador to the U.S., Prince Bandar bin Sultan, in Washington D.C. was elusive as to whether the project proposals—expected to be of "direct benefit to the kingdom" (Robins 2004, 324)—concerned the oil upstream sector as well. Saudi Aramco, supported by Al-Naimi, managed to limit the initiative to gas only and made sure that the upstream gas ventures were not agreed upon until the terms were favourable to Aramco which took its stake in all of them. The Saudi Gas Initiative was mentioned as an epitome of Saudi Aramco's clout by four interviewees (think-tank, October 7, 2011; government, February 14, 2012; industry, December 13, 2011 and March 15, 2012) and also, by the participants in the "Gulf Energy Challenges" workshop at the Gulf Research Meeting 2012 at Cambridge University, where this chapter was initially presented in a draft form.

11 In this respect, van der Meulen's comprehensive work on the UAE's tribal history also shows that "the Al-Nuhayyan shaykhs have had uncontested leadership over most of modern-day Abu Dhabi emirate since at least the late 1700s" (1997, 21). Please note that different transliterations are used for the same name: Al-Nahyan, Al-Nuhayyan, Al-Nahayan.

12 For a list of the Bani Yas tribal sections, see van der Meulen (1997, 85).

13 It is worth noting that the history of Abu Dhabi is marked both by internecine conflict for rule in the Al-Nahyan family for over one century and by external threats to territorial integrity such as the Wahhabis of the interior Arabia, the Qawasim tribe of Sharjah and Ra's al-Khaimah, and British imperialism. For a well-documented history of Abu Dhabi, see Davidson (2009a).

14 PDTC was formed in 1935 through the partnership of Anglo-Persian (now BP), Shell, Compagnie Francaise des Petroles (CFP, now Total), Exxon and Mobil (Rai and Victor 2012, 482).

15 This was co-owned by the Anglo-Persian Company (now BP) by two thirds and Companie Francaise des Petroles (now Total) by one third (Suleiman 2007a, 25ff.).

16 In 1961, the ruler even installed a ban on all new construction work in Abu Dhabi, which stayed in place until Sheikh Zayed the Second took upon the leadership in 1966 by deposing his brother with the support of the Al-Nahyan and the British (Davidson 2009a, 33f.). Ironically, the smaller sheikhdoms of the lower Gulf benefited from Abu Dhabi's oil wealth more than the oil producing sheikhdom itself as Sheikh Shakhbut was obliged to give them 4 percent of the oil revenues in conformity with a clause from the British oil concession (Davidson 2007, 35).

17 "The people of Abu Dhabi could see them importing cars, trucks, materials and equipment but since nothing was being bought here, there was no benefit to the local economy. The local merchants could not provide the needed products and services nor could they accommodate the increasing number of oil workers

because they lacked capital and were forbidden from building anything" (Al Fahim 1995, 94).

18 One of the interviewees (academia, February 10, 2012) cared to emphasize this.

19 These are similar to ministries with the observation that they were devolved to departments as soon as the federal government of the United Arab Emirates was created in 1971. While the structure of the political system changed with the introduction of federal-level ministries, the members of the ruling family continued to dominate the new state's posts. For this, see Figure 5.13 on the state posts in the mid-1990s in Herb (1999, 138).

20 For a well-informed discussion of the role of each of these tribes and the functions they occupy in Abu Dhabi, see van der Meulen (1997), Chapter 3, "Tribal and Kinship Ties in Abu Dhabi Emirate Politics."

21 The details about the creation of the federation, the rounds of negotiations, constitutional crisis, historical divides, economic and demographic imbalances, the stance of Qatar and Bahrain on joining the federation, the hesitation of Ra's Al-Khaimah until early 1972, the role of different personalities in the foundation of the UAE and the institutional set-up go beyond the scope of this chapter. For this, there are some excellent records such as Davidson (2005), and Heard-Bey and Lange (2010).

22 In order to bring the emirates together in a federation, Sheikh Zayed understood that the autonomy of each of the seven emirates and their respective ruling families had to be observed. They were thus able to preserve the old patriarchal structures; this is why, to the present, local administration and political loyalties continue to follow tribal affiliation (Heard-Bey and Lange 2010; Koch 2011; van der Meulen 1997). At the federal level, despite Abu Dhabi's substantial contribution to the federal fund, the Ruler of Abu Dhabi—though at the same time, President of the UAE—has cared primarily for his local turf and only secondarily for the other emirates in the federation (Heard-Bey 2005, 338).

23 This is mainly due to the comparative advantage of cheap feedstock.

24 For a list of Abu Dhabi's SWFs and detailed information, see Davidson (2009c), and *The Economist*, "Special Report, State Capitalism—The Visible Hand" (January 21, 2012). The value of these SWFs, all of them government-owned, is estimated to be over $1 trillion and among the largest globally, far ahead of Singapore, Norway or other Gulf investors like Kuwait, Dubai or Qatar (Davidson 2009c, 62).

25 The UAE has the poorest record on all of these three forms of freedom in the region (Koch 2011).

26 FNC is a quasi-parliament with consultative role only.

27 The reason for this acceleration may be the need of the ruling family to uphold increasing government largesse in the coming years, which indigenous citizens have taken for a fact (*The Economist*, 24 November 2012).

28 For a list of these concessions, see Suleiman (1988).

29 For this, see Davidson (2009c) and Krane (2010, 2012).

30 This is in fact not the task of the case studies, but instead, of the statistical analysis—see section 4.4.

References

Abu-Nasr, Donna, "Saudis Skip Arab Spring as Nation Pours Money into Jobs," in: *Bloomberg Markets Magazine* (online), April 2, 2012, available at www.bloomberg.com/news/2012-04-02/saudis-skip-arab-spring-as-nation-pours-money-into-jobs.html (status: February 28, 2013).

ADMA-OPCO, *Official Website*, www.adma-opco.com/(status: February 28, 2013).

ADNOC, Official Website, www.adnoc.ae/(status: March 1, 2013).

Al-Attar, Abdulaziz, and Osamah Alomair, "Evaluation of Upstream Petroleum Agreements and Exploration and Production Costs," in: *OPEC Review*, 29 (2005), 243–266.

Al Fahim, Mohammed, *From Rags to Riches* (London Centre of Arab Studies, London, 1995).

Al-Rasheed, Madawi, "Circles of Power: Royals and Society in Saudi Arabia," in: *Saudi Arabia in the Balance: Political Economy, Society, Foreign Affairs* (Hurst and Company, London, 2005), 185–213.

Al Sadik, Ali Tawfi, "Evolution and Performance of the UAE Economy 1972–1998," in: *United Arab Emirates: A New Perspective* (Trident Press, London, 2001), 202–230.

Ammoun, Camille, "The Institutionalization of the Saudi Political System and the Birth of 'Political Personnel,'" in: *Constitutional Reform and Political Participation in the Gulf* (Gulf Research Center, Dubai, 2006), 211–238.

Audinet, Pierre, Paul Stevens, and Shane Streifel, *Investing in Oil in the Middle East and North Africa. Institutions, Incentives and the National Oil Companies* (World Bank, Sustainable Development Department, Middle East and North Africa Region, Washington DC, 2007).

Baqi, Mahmoud M. Abdul and Nansen G. Saleri, Fifty-Year Crude Oil Supply Scenarios: *Saudi Aramco's Perspective*, Saudi Aramco Presentation, February, 24 2004, CSIS, Washington D.C., available at www.nog.se/files/040224_baqiand saleri.pdf (status: February 28 2013).

BBC, *Arab uprising: Country by country–Saudi Arabia*, August 31, 2012, available at www.bbc.co.uk/news/world-12482678 (status: February 27, 2013).

Biernacki, Patrick and Dan Waldorf, "Snowball sampling: problem and techniques of chain referral sampling," in: *Sociological Methods and Research*, 10 (1981), 141–163.

Blight, Garry, Sheila Pulham and Paul Torpey, "Arab spring: an interactive timeline of Middle East protests," in: *The Guardian*, January 5, 2012, available at www.guardian.co.uk/world/interactive/2011/mar/22/middle-east-protest-interactive-timeline?CMP=twt_gu (status: February 28, 2013).

Bogner, Alexander, and Wolfgang Menz, „Das Theoriegenerierende Experteninterview. Erkenntnisinteresse, Wissensformen, Interaktion," in: *Das Experteninterview. Theorie, Methode, Anwendung*, 2 (2005), 33–70.

—, „Experteninterviews in der Qualitativen Sozialforschung. Zur Einführung in eine sich intensivierende Methodendebatte," in: *Das Experteninterview*, 3 (2009), 7–31.

British Petroleum, *BP Statistical Review of World Energy*, June 2011, available at www.bp.com/assets/bp_internet/globalbp/globalbp_uk_english/reports_and_publications/statistical_energy_review_2011/STAGING/local_assets/pdf/statistical_review_of_world_energy_full_report_2011.pdf (status: February 27, 2013).

Bronson, Rachel, *Thicker than Oil: America's Uneasy Partnership with Saudi Arabia* (Oxford University Press, New York, 2006).

Brumberg, Daniel, "The Trap of Liberalized Autocracy," in: *Journal of Democracy*, 13 (2002), 56–68.

Butt, Gerald, "Oil and Gas in the UAE," in: *United Arab Emirates. A New Perspective* (Bookcraft, Stroud, 2001), 231–248.

Chaudhry, Kiren Aziz, "The Price of Wealth: Business and State in Labor Remittance and Oil Economies," in: *International Organization*, 43 (1989), 101–145.

—, "Economic Liberalization and the Lineages of the Rentier State," in: *Comparative Politics*, 27/1 (1994), 1–25.

CIA The World Factbook, *Saudi Arabia*, available at www.cia.gov/library/publications/the-world-factbook/geos/sa.html (status: September 18, 2012).

—, *United Arab Emirates*, www.cia.gov/library/publications/the-world-factbook/geos/ae.html (status: September 18, 2012).

Cordesman, Anthony H., *Saudi Arabia Enters the Twenty-first Century: The Political, Foreign Policy, Economic, and Energy Dimensions* (Greenwood Publishing Group, Westport, CT, 2003).

—, *Security Cooperation in the Middle East* (Center for Strategic and International Studies, Washington DC, 2007).

—, *Security Challenges and Threats in the Gulf: A Net Assessment* (Center for Strategic and International Studies, Washington DC, 2008).

Crystal, Jill, "Gulf Security in the 1990s: Review Article," in: *The Middle East Journal*, 52/2 (1998), 277–280.

Davidson, Christopher, *The United Arab Emirates: A Study in Survival* (Lynne Rienner, Boulder, CO, 2005).

—, "The Emirates of Abu Dhabi and Dubai: Contrasting Roles in the International System," in: *Asian Affairs*, 38 (2007), 33–48.

—, *Abu Dhabi: Oil and Beyond* (Columbia University Press, New York, 2009a).

—, "The United Arab Emirates: Prospects for Political Reform," in: *Brown Journal of World Affairs*, 15 (2009b), 117–127.

—, "Abu Dhabi's New Economy: Oil, Investment and Domestic Development," in: *Middle East Policy*, 16 (2009c), 59–79.

—, *The United Arab Emirates: Frontiers of the Arab Spring*, September 8, 2012, available at www.opendemocracy.net/christopher-m-davidson/united-arab-emirates-frontiers-of-arab-spring (status: February 28, 2013).

Dexter, Lewis Anthony, *Elite and Specialized Interviewing* (European Consortium for Political Research, Amsterdam, 2006).

DiPaola, Anthony, Abu Dhabi to Operate Onshore Oil Fields as Concessions Expire, on Bloomberg, January 8, 2014, available at www.bloomberg.com/news/2014-01-08/abu-dhabi-to-operate-onshore-oil-fields-as-concessions-expire.html (last accessed on February 12, 2014).

Dunning, Thad, "Resource Dependence, Economic Performance, and Political Stability," in: *Journal of Conflict Resolution*, 49 (2005), 451–482.

The Economist, Special Report: State Capitalism – The Visible Hand, January 21, 2012, available at www.economist.com/node/21542924 (status: February 28, 2013).

—, *Where are the jobs for boys?*, November 24, 2012, available at www.economist.com/news/middle-east-and-africa/21567128-recent-government-efforts-create-jobs-may-store-up-trouble-future-where (status: February 27, 2013).

Ehteshami, Anoushiravan, and Emma C. Murphy, "Transformation of the Corporatist State in the Middle East," in: *Third World Quarterly*, 17 (1996), 753–772.

Ehteshami, Anoushiravan, and Steven Wright, "Political Change in the Arab Oil Monarchies: From Liberalization to Enfranchisement," in: *International Affairs*, 83 (2007), 913–932.

EIA (Energy Information Administration), *Country Analysis Brief: Saudi Arabia*, 2013a, available at www.eia.gov/countries/analysisbriefs/Saudi_Arabia/saudi_arabia.pdf (status: February 28, 2013).

—, *Country Analysis Brief: United Arab Emirates*, 2013b, available at www.eia.gov/countries/analysisbriefs/UAE/uae.pdf (status: February 28, 2013).

Field, Michael, *The Merchants: The Big Business Families of Arabia* (J. Murray, London, 1984).

Gause, Gregory F., *The International Relations of the Persian Gulf* (Cambridge University Press, Cambridge, 2010).

Gerring, John, "What Is a Case Study and What Is It Good For?" in: *American Political Science Review*, 98 (2004), 341–354.

—, "Is There a (viable) Crucial-case Method?" in: *Comparative Political Studies*, 40 (2007), 231–253.

Glosemeyer, Iris, "Checks, Balances and Transformation in the Saudi Political System," in *Saudi Arabia in the Balance: Political Economy, Society, Foreign Affairs* (Hurst and Company, London, 2005), 214–233.

Halliday, Fred, *Nation and Religion in the Middle East* (Saqi Books, London, 2000).

Heard-Bey, Frauke, "The United Arab Emirates: Statehood and Nation-building in a Traditional Society," in: *The Middle East Journal*, 59 (2005), 357–375.

Heard-Bey, Frauke, and Wigand Lange, *Die Vereinigten Arabischen Emirate zwischen Vorgestern und Übermorgen* (Olms, New York, 2010).

Herb, Michael, *All in the Family: Absolutism, Revolution, and Democracy in the Middle Eastern Monarchies* (SUNY Press, New York, 1999).

Herb, Michael, "Princes and Parliaments in the Arab World," in: *The Middle East Journal*, 58 (2004), 367–384.

Hertog, Steffen, "Building the Body Politic. Emerging Corporatism in Saudi Arabia," in: *Chroniques yéménites*, 12 (2004), available at http://cy.revues.org/187 (status: March 1, 2013).

—, "Segmented Clientelism: The Political Economy of Saudi Economic Reform Efforts," in: *Saudi Arabia in the Balance: Political Economy, Society, Foreign Affairs* (Hurst and Company, London, 2005), 111–143.

—, "The New Corporatism in Saudi Arabia: Limits of Formal Politics," in: *Constitutional Reform and Political Participation in the Gulf* (Gulf Research Center, Dubai, 2006), 241–275.

—, "Shaping the Saudi State: Human Agency's Shifting Role in Rentier-State Formation," in: *International Journal of Middle East Studies*, 39 (2007), 539–563.

—, "Petromin: The Slow Death of Statist Oil Development in Saudi Arabia," in: *Business History*, 50 (2008), 645–667.

—, *Princes, Brokers, and Bureaucrats: Oil and the State in Saudi Arabia* (Cornell University Press, New York, 2010a).

—, "Defying the Resource Curse: Explaining Successful State-Owned Enterprises in Rentier States," in: *World Politics*, 62 (2010b), 261–301.

Hourani, Albert, *A History of the Arab Peoples* (Harvard University Press, Cambridge, MA, 2010).

International Labour Office (ILO), *Global Employment Trends 2011. The Challenge of a Jobs Recovery*, Geneva, 2011, available at www.ilo.org/wcmsp5/groups/public/@dgreports/@dcomm/@publ/documents/publication/wcms_150440.pdf (status: February 28, 2013).

Jadwa Investment, *Saudi Arabia's Coming Oil and Fiscal Challenge*, July 2011, www.jadwa.com/en (status: February 28, 2013).

Kechichian, Joseph A., *Power and Succession in Arab Monarchies: A Reference Guide* (Lynne Rienner, Boulder, CO, 2008).

Kobayashi, Yoshikazu, "Corporate Strategies of Saudi Aramco," *The James A. Baker III Institute for Public Policy – Rice University Working Papers*, 2007, available at www.bakerinstitute.org/programs/energy-forum/publications/energy-studies/docs/NOCs/Papers/NOC_Kobayashi%20SAramco.pdf (status: February 22, 2013).

Koch, Christian, "Economics Trumps Politics in the United Arab Emirates," in: *Political Change in the Arab Gulf States: Stuck in Transition* (Lynne Rienner, Boulder, CO, 2011), 167–189.

Krane, Jim, "Energy Conservation Options for GCC Governments," *Dubai School of Government Policy Brief*, 18 (2010), available at www.jimkrane.com/misc/DSG_Policy_Brief_18_English-FINAL.pdf (status: February 22, 2013).

—, "Energy Policy in the Gulf Arab States: Shortage and Reform in the World's Storehouse of Energy," in: *2012 USAEE Working Papers*, available at www.usaee.org/usaee2012/submissions/OnlineProceedings/KRANE_IAEE_Energy-Policy-in-the-Gulf_Sept2012.pdf (status: February 22, 2013).

Legard, Robin, Jill Keegan, and Kit Ward, "In-depth Interviews," in: *Qualitative Research Practice: A Guide for Social Science Students and Researchers* (Sage Publications, London, 2003), 138–169.

Long, David E., and Sebastian Maisel, *The Kingdom of Saudi Arabia* (University Press of Florida, Gainesville, FL, 2010).

Luciani, Giacomo, "Allocation vs. Production States: A Theoretical Framework," in: *The Arab State* (University of California Press, Berkeley, CA, 1990), 65–84.

—, "From Private Sector to National Bourgeoisie: Saudi Arabian Business," in: *Saudi Arabia in the Balance: Political Economy, Society, Foreign Affairs* (Hurst and Company, London, 2005), 144–181.

Marcel, Valerie, and John V. Mitchell, "How It All Started," in: Marcel, Valerie, *Oil Titans: National oil companies in the Middle East* (Brookings Institution Press, Washington, D.C., 2006), pp. 14–36.

Mason, Jennifer, *Qualitative Researching* (Sage Publications, London, 2002).

Mazzetti, Mark and Emily B. Hager, *Secret Desert Force Set Up by Blackwater's Founder*, New York Times, May 14, 2011, available at www.nytimes.com/2011/05/15/world/middleeast/15prince.html?pagewanted=all&_r=0 (status: February 28, 2013).

van der Meulen, Hendrik, *The Role of Tribal and Kinship Ties in the Politics of the United Arab Emirates*, PhD Thesis, The Fletcher School of Law and Diplomacy, May 1997.

Meuser, Michael, and Ulrike Nagel, "The Expert Interview and Changes in Knowledge Production," in: *Interviewing Experts* (Palgrave Macmillan, London, 2009), 17–42.

Millwood, Joanna, and M. Robin Heath, "Food Choice by Older People: The Use of Semi-Structured Interviews with Open and Closed Questions," in: *Gerodontology*, 17 (2008), 25–32.

Myers Jaffe, Amy, and Jareer Elass, "Saudi Aramco: National Flagship with Global Responsibilities," *The James A. Baker III Institute for Public Policy – Rice*

University Working Papers, 2007, available at www.bakerinstitute.org/programs/
energy-forum/publications/docs/NOCs/Papers/NOC_SaudiAramco_Jaffe-Elass-
revised.pdf (status: February 22, 2013).

Niblock, Tim, *State, Society, and Economy in Saudi Arabia* (Croom Helm, London,
1982).

Niblock, Tim, and Monica Malik, *The Political Economy of Saudi Arabia*
(Routledge, London, 2007).

Oil and Gas Journal, Old onshore concession ends in Abu Dhabi, January 22, 2014,
available at www.ogj.com/articles/2014/01/old-onshore-concession-ends-in-abu-
dhabi.html (last accessed on February 12, 2014).

Okruhlik, Gwenn, "Rentier Wealth, Unruly Law, and the Rise of Opposition: The
Political Economy of Oil States," in: *Comparative Politics*, 31/3 (1999),
295–315.

PFC Energy, *Forces Shaping Strategic Themes and Hubs*, Presentation in Washington
D.C., February 10–11, 2011.

Rai, Varun, and David G. Victor, "Awakening Giant: Strategy and Performance of
the Abu Dhabi National Oil Company (ADNOC)," in: *Oil and Governance.
State-Owned Enterprises and the World Energy Supply* (Cambridge University
Press, New York, 2012), 478–514.

Robins, Philip, "Slow, Slow, Quick, Quick, Slow: Saudi Arabia's 'Gas Initiative,'" in:
Journal of Energy Policy, 32 (2004), 321–333.

Ross, Michael, *The Oil Curse: How Petroleum Wealth Shapes the Development of
Nations* (Princeton University Press, Princeton, NJ, 2012).

Salehi-Isfahani, Djavad, "Human Development in the Middle East and North
Africa," in: *UNDP Human Development Research Paper*, 26 (2010), available
at http://hdr.undp.org/sites/default/files/HDRP_2010_26.pdf (status: March 11,
2014).

Saudi Aramco, *Official Website: Our Company – People*, www.saudiaramco.com/
en/home.html#our-company%257C%252Fen%252Fhome%252Four-
company%252FOur-people-demonstration.baseajax.html (status: February 28,
2013).

Saudi Gazette, *Saudi Aramco Board Reconstituted*, August 25, 2010, available at
http://saudigazette.com.sa/index.cfm?method=home.regcon&contentID=
2010082281613 (status: February 28, 2013).

Seawright, Jason, and John Gerring, "Case Selection Techniques in Case Study
Research: A Menu of Qualitative and Quantitative Options," in: *Political Research
Quarterly*, 61 (2008), 294–308.

Seznec, Jean-François, and Mimi Kirk, *Industrialization in the Gulf: A Socioeconomic
Revolution* (Taylor & Francis, London, 2010).

Shihab, Mohamed, "Economic Development in the UAE," in: *United Arab Emirates:
A New Perspective* (Trident Press, London, 2001), 249–259.

Simmons, Matthew R., *Twilight in the Desert: The Fading Of Saudi Arabia's Oil*,
Presentation at Hudson Institute in Washington D.C., September 9, 2004,
available at http://sepwww.stanford.edu/sep/jon/world-oil.dir/stork3.pdf (status:
February 28, 2013).

—, *Twilight in the Desert: The Coming Saudi Oil Shock and the World Economy*
(John Wiley & Sons, Hoboken, NJ, 2005).

Sorenson, David S., *An Introduction to the Modern Middle East: History, Religion,
Political Economy* (Westview Press, Boulder, CO, 2007).

Steinberg, Guido, "The Wahhabi Ulama and the Saudi State: 1745 to the Present," in: *Saudi Arabia in the Balance: Political Economy, Society, Foreign Affairs* (Hurst and Company, London, 2005), 11–34.

Stevens, Paul, "National Oil Companies and International Oil Companies in the Middle East: Under the Shadow of Government and the Resource Nationalism Cycle," in: *The Journal of World Energy Law & Business*, 1 (2008), 5–30.

—, "Saudi Aramco: The Jewel in the Crown," in: *Oil and Governance. State-Owned Enterprises and the World Energy Supply* (Cambridge University Press, New York, 2012), 173–233.

Suleiman, Atef, "Oil Experience of the United Arab Emirates and Its Legal Framework," in: *Journal of Energy & Natural Resources Law*, 6 (1988), 1–24.

—, *The Petroleum Experience of Abu Dhabi* (Emirates Center for Strategic Studies, Abu Dhabi, 2007a).

—, "The United Arab Emirates," in: *Encyclopedia of Hydrocarbons* (Istituto de la Enciclopedia Italiana, Roma, 2007b), Vol. IV: Hydrocarbons: Economics, Policies and Legislation, 773–784.

Tétreault, M. A., "The Political Economy of Middle Eastern Oil," in: *Understanding the Contemporary Middle East* (Lynne Rienner, Boulder, CO, 2008), 255–279.

Thurber, Mark C., David R. Hults, and Patrick Heller, "Exporting the 'Norwegian Model': The Effect of Administrative Design on Oil Sector Performance," in: *Journal of Energy Policy*, 39 (2011), 5366–5378.

Tordo, Silvana, Brandon S. Tracy, and Noora Arfaa, *National Oil Companies and Value Creation* (World Bank Publications, Washington DC, 2011).

Ulrichsen, Kristian, *Gulf Security: Changing Internal and External Dynamics* (Kuwait Programme on Development, Governance and Globalisation in the Gulf States, LSE Publications, 2009a).

—, "Internal and External Security in the Arab Gulf States," in: *Middle East Policy*, 16 (2009b), 39–58.

Vitalis, Robert, *America's Kingdom: Mythmaking on the Saudi Oil Frontier* (Stanford University Press, Redwood City, CA, 2006).

Yergin, Daniel, *The Prize: The Epic Quest for Oil, Money and Power* (Free Press, New York, 1991).

—, *The Prize: The Epic Quest for Oil, Money and Power*, with a new epilogue (Free Press, New York, 2009).

6   Conclusions

This concluding chapter first compares and contrasts the results of the statistical analysis with the lessons drawn from the two case studies. Based on this, the second part revises the analytical framework and enhances it with the aim of explaining oil upstream policy choices globally in a more comprehensive manner. The third and final part is devoted to remarks about the relevance and limitations of this book, proposes a number of research ways forward, and not least, discusses the policy implications of this research.

6.1 Comparative results of the empirical analysis

As part of the mixed-methods design, the quantitative analysis and the case studies reinforce and complement each other. In summary, the statistical analysis renders the domestic constraints of politico-institutional and economic nature—namely, executive constraints, oil reliance and the interaction of the two—highly relevant for the oil upstream policy choice. The contextual factors in the statistical setup achieve significance across models only as concerns the geological conditions. The national oil company (NOC) technical capabilities fail to reach the statistical significance threshold whereas the oil price variable generates the opposite sign to the one expected in the analytical framework (see discussion in section 4.4).

By comparison, the case studies confirm the role of political constraints, oil reliance and also geology, find no explanatory power for oil prices, and—opposite to the statistical results—render the technical capabilities of the state-owned oil company quite important for the oil upstream policy choice. The case studies further uncover four other major explanations: the historical premises of the modern state; the oil production start as an indicator of the actual beginnings of the oil industry in the respective state; demographics as a proxy both for popular pressures for nationalization and/or the need for diversification of security; and not least, the proclivity of the ruling family and ruling elites to efficient oil upstream policy choices in the absence of executive constraints (see section 5.5).

The case studies also reveal that apart from geology and oil prices, the other three factors included in the analytical framework—that is, technical capabilities of the NOC, oil reliance and political constraints—are heavily embedded in the historical path of development. The historical path dependency which proves to be a stronghold of upstream sector policy choices in the narrative of the case studies has been carefully accounted for in the statistical analysis by using the dynamic panel fractional (DPF) estimator in the Tobit. Given that the inclusion of a lag of the dependent variable is quite disputed among statisticians (see section 4.3), the empirical evidence of the case studies suggests that its omission might have been more problematic. Therefore, the findings of the case studies reinforce the choice of the statistical method in the quantitative analysis.

Furthermore, the statistical analysis finds that the interaction between executive constraints and oil reliance is consistently significant at high confidence levels across all models which test for it. The effect of the interaction is straightforward based on the statistical analysis, yet the case studies bring more fine-tuning to the effect of limited executive constraints and high oil reliance on the oil upstream sector policy. More specifically, the case studies provide insights into how the absence of political constraints makes the incumbents more prone to pay consideration to the contextual factors because they do not perceive the threat of being ousted as imminent as they otherwise would. While political constraints in the statistical analysis can capture the room for policy change at the disposal of the ruling elites, the case studies go one step further and flesh out how in the absence of executive constraints, the role of the ruling family becomes instrumental in shaping the political economy of the oil sector, building institutions that follow family structure instead of function, designing allocative rather than productive states, and yet making efficient decisions as to the involvement of international oil companies (IOCs) in the oil upstream sector. Along these lines, the case studies also offer the reasoning behind the relatively high, negative correlation between executive constraints and oil reliance proxied by oil rents of GDP and oil exports of GDP (see section 4.2.2 and correlation matrix in Annex 6). The two case studies show how, if their room for maneuver permits, ruling elites may intentionally pursue high reliance on the oil sector as a politically rational strategy to remain in power.

All in all, the case studies are complementary and reinforcing with respect to the quantitative analysis. While the latter offers crisp results and cannot provide the argument behind a given variable (e.g. GDP per capita may account for different explanations), the former shed light on the explanations behind statistical relations and identify aspects that are difficult to quantify (e.g. diversification of security). At the same time, the case studies can neither offer large-N generalizations of the relations nor tell which arguments may have higher explanatory power for the oil upstream sector policy. In turn, the statistical analysis can determine the general trends across the population of oil producing countries and prove specific variables consistently significant across several different statistical models.

6.2 Revised analytical framework

Based on the combined findings from the quantitative analysis and case studies, the analytical framework proposed in section 3.2 needs revisions.

These revisions are not major as the foundations of the analytical framework remain the same. The context is changed in the sense that one factor is taken out while three others are newly added. Also, the effect of the interaction between the two main explanatory variables is pinned down.

The context is framed by five factors, which fall under three types of explanations: technical, historical and security-related explanations. The technical factors remain unchanged, as in the initially proposed framework—namely, geology and NOC's technical capabilities—despite the fact that the quantitative analysis has not rendered the latter statistically significant. As the case studies have shown, the technical expertise of the NOC plays an important role in explaining oil upstream policy choices, especially when the limits on the executive are low and the ruling elites are

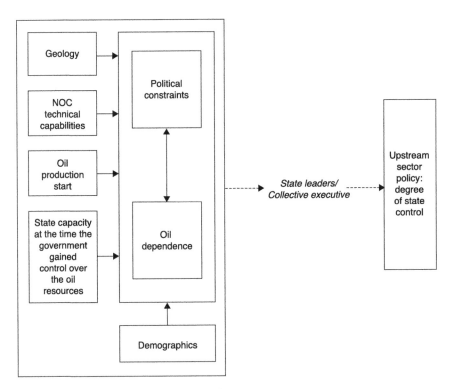

Figure 6.1 Revised analytical framework

more likely to take economically rational decisions in the oil upstream sector. Due to data shortage, no better proxies for the NOC technical capabilities could be factored into the statistical analysis, yet the empirical evidence from the two case studies clearly speaks for the leverage of the NOC and its technical capacity. The effects of both technical variables stay the same as postulated in the initial analytical framework.

The contextual factors of historical nature are: the oil production start and the state capacity at the time of nationalizations or when the government gained control over the oil resources. More explicitly, the start of the oil production is an indicator of the beginning of the oil industry in the respective country. The further back in time this lies, the more likely it is that the oil industry has acquired experience and, thereby, a certain level of technical expertise and human capital. The later the oil production starts, the higher the foreign control in the upstream sector is likely to be. Producer states with young oil industries are in need of foreign assistance to bring their oil production up to speed and reach a certain level.

Also of historical nature, the state capacity at the time the government gained control over the oil resources is highly relevant in explaining oil upstream policy choices. State capacity is a good indicator of the government's capabilities to run the oil sector through the NOC if that chance arose (such as with the wave of nationalizations or gaining independence in the post-Soviet space). Furthermore, it can account for the position and ability of the state when bargaining with foreign oil firms—for example, whether the former has enough leverage to coopt the latter on less favorable terms (e.g. under service contracts) or whether the former has a limited bargaining power and needs to make compromises and give in control rights to be able to attract foreign expertise and uphold oil production. Consequently, the lower the state capacity was at the time the government gained control over the oil resources, the more foreign control is expected in the oil upstream sector.

Both historical factors—oil production start and state capacity at the time of gaining control—together with the security-related explanation (i.e. demographics) are inspired by the results of the case studies. In this revised framework, demographics represent a way of projecting power by a state. A small state is more likely to tie in foreign operators into the upstream sector in order to increase its national security. This is why, the lower the demographics are, the higher the state's need for diversification of security becomes and thereby, the more foreign control is expected in the upstream sector.

The context, shaped by the five factors, creates the basis for policy-making in the oil upstream sector. The higher the political constraints are, the less room for maneuver the ruling elites have and consequently, the closer the oil upstream sector policy choice comes to the public preference (i.e. state control of the national oil wealth). In short: the higher the executive constraints are, the more state control is expected in the oil upstream sector.

The degree of national oil reliance is only relevant in explaining oil upstream policy choice in interaction with low executive constraints. In the absence of political constraints, the incumbents do not feel the threat of being ousted from power and thus do not need to play the card of political rationality. Low executive constraints are likely to make the incumbents more attentive to the contextual conditions, increasingly so when the state leader and the executive are heavily reliant on the money flows from the oil sector—that is, when the national oil reliance is high. The more efficiently the oil upstream industry is run, the more money flows into the coffers of the state. On these premises, the incumbents are more prone to take note of the contextual parameters and make informed and economically rational decisions regarding foreign expertise and the share-out of control rights. While the two case studies uncovered that the ruling elites intentionally nurtured and have upheld high oil reliance as a strategy to stay in power, this does not need to be the rule in oil producing countries. It can more generally be the result of the economic structures in place. The important take-away is that the likelihood of the ruling elites to adopt economically rational upstream sector policies—that is, to decide based on the contextual parameters—increases with lower executive constraints and higher national oil reliance.

All in all, this revised analytical framework has drawn on the initial framework proposed in section 3.2. The extent of executive constraints is key to understanding the outcome on the oil upstream sector policy. When political constraints are high, the likely outcome is a rather closed upstream sector—largely irrespective of the contextual factors on the ground. In turn, when executive constraints are low, the incumbents are more prone to guide their policy decisions in the oil upstream sector based on technical factors, historical elements which impact the status of the oil industry and, not least, on national security considerations. This proclivity to more economically efficient oil upstream policies—given the contextual parameters in place—is increasing with a higher extent of national oil reliance.

6.3 Final remarks

This book started off from the question: *Why have oil producing countries pursued diverse policies in their oil upstream sector?* The empirical evidence spanning producer countries all across the world, from Latin America over the post-Soviet space to the Middle East and North Africa, displays large variation in the policies adopted in the oil upstream industry. While the corporate finance literature abounds in studies of firm control, there is hardly any research on the determinants of industry or sectoral control. To fill this niche, this book put forward an analytical framework which explains industry control structures in the oil upstream sector. After a careful scrutiny of different strands of literature for potential trigger factors of oil upstream sector policies, a rough analytical framework was proposed.

This was then tested in a mixed-methods design based on both primary and secondary data. While a large-N analysis uncovered the general trends of how oil upstream sector policies are chosen across oil producing countries worldwide, two case studies from the Middle East brought insights into more nuanced and less quantifiable explanations. Based on the combined results of the quantitative and qualitative analyses, the analytical framework was refined and enhanced to stand the test of further analyses of oil upstream sector policies worldwide.

Given the importance of petroleum in general, and the producer countries in particular, to energy security worldwide, understanding government control is important. Indeed, one of the first books on the industry written after the energy crisis of 1973/1974 is titled *The Control of Oil* (Blair 1976), which also inspired the title for the present book. However, Blair's work, similarly to other books and articles, tends to focus on geopolitics and economics. In contrast, this book is comparative, seeking to explain variation in government policies toward the domestic oil sector. By exploiting differences in domestic environments to understand variations in domestic government policies, this work falls nicely into Comparative Politics.

There are several contributions that this book tried to make both theoretically and empirically. Yet as any piece of research, it also has its flaws. Both are recounted in the following, before any new research ideas are postulated.

6.3.1 Contributions, limitations, and research ways forward

To start with the *contributions*, these are mainly fourfold. *First*, this book conceptualized and operationalized one type of economic policy which has been marginally addressed so far: upstream sector policy. To this end, the concepts of ownership and control at the level of the oil production industry have been disentangled and three basic schemes for upstream sector policies—(1) state ownership with state control; (2) state ownership with private/mixed control; and (3) private ownership with private control—have been proposed. By using control structures as a lens for viewing variations in oil sector policy, this book enriches the way research views the question and proposes a new focus in the field of International Political Economy (IPE)—namely, sectoral or industry control.

Second, to explain variation in oil upstream sector policies, this book bridged IPE and Political Science while bringing together several bodies of literature, i.e. the Nationalization/Expropriation scholarship, NOCs literature, Resource Curse literature, Comparative Politics, and Middle East Area Studies. For technical aspects related to the subject of this research, engineering studies were also reviewed and several lessons applied. An interdisciplinary analytical framework was further advanced to explain upstream sector policy choices in oil producing countries worldwide. This is the substantial theoretical added value of this book to the existing literature.

While the initially proposed framework was the product of a theory development exercise, the revised form (in section 6.2) draws on extensive empirical evidence and data analysis.

In summary, this analytical framework portends that the extent of political constraints in the oil producing country is crucial to explicate the policy choice in the upstream. The higher the level of constraints is, the more likely it is that the policy choice converges to the public preference—namely, high/complete state control in the oil upstream—regardless of the contextual parameters on the ground which might shape requirements not attainable by the NOC. In turn, the lower the limits on the policy-making are, the more prone the incumbents become to rational upstream policies from an economic perspective in view of the context-defining factors. These are geological conditions, technical capabilities of the NOC, oil production start, state capacity at the time of gaining control over the oil resources, and demographics. This inclination towards economic rationality increases with higher degree of oil dependence of the country in question. Finally, this framework shows that policies in the upstream sector, which elicits the highest revenues along the petroleum value supply chain, are not determined exclusively by one type of factors such as technical, economic or political, but by a combination thereof.

Third, for the purposes of the quantitative analysis, a dataset was compiled from eight different sources, while five variables were self-coded. To put together such a database, a lot of time, effort and not least, financial resources for the data on the dependent variable had to be invested. This book puts data at the disposal of other scholars. Though in an aggregate form, this is still a good starting point for further analysis.

Fourth, a wealth of primary data on the political economy of the petroleum sector, operating oil companies, relevant decision-makers and other stakeholders in the oil industry of Saudi Arabia and Abu Dhabi/the UAE has been collected through interviews with 33 experts at high official level and from various domains of activity. Given that such information is quite insightful and not publicly available elsewhere, the two case studies build a significant empirical contribution to the existing scholarship. At the same time, this type of information may inform the work of policy-makers, oil market analysts and industry professionals focusing on the two extremely oil-rich producers from the Middle East region.

As any other scholarly work, this book also has its *shortcomings* which are related particularly to data issues. *First*, the quantitative analysis was limited to the period 1987 to 2010 due to the lack of existing data on oil production by companies going further back in time. However, despite the confined time scope, the statistical analysis allowed for the identification of general trends across the population of oil producing countries worldwide.

Second, more case studies from other parts of the world like Latin America or Africa might have brought more insights into potential regional differences and also, more fine-tuning to the revised analytical framework explaining global determinants of oil upstream policies. Yet time constraints did not allow for more extensive research than what had already been done. Contacting so many experts and interviewing on a relatively sensitive topic can be an enormous hurdle especially when the community of experts is quite compact and not always accessible, as was the case with Middle East energy experts.

Third, the quantitative analysis did not find support for one explanation for which the case studies raised extensive evidence—namely, the role of the NOC's technical capabilities. Due to the fact that this type of data is proprietary, proxies had to be used, which (arguably) did not manage to capture the argument accurately. In the presence of better data, the statistical results are very likely to have conformed to the findings of the case studies.

Fourth, the revised analytical framework put forward a number of explanations like state capacity at the time the government in the oil producing country gained control over the oil resources or demographics, for which data might be difficult to find, especially if the quantitative analysis goes back to the 1970s. Nonetheless, these are all issues that scholars and practitioners alike need to continue struggling with.

On these self-reflective premises, a number of research ways forward can be envisioned. Along the lines of this book, international political economists, political scientists as well as historians are strongly encouraged to expand the existing scholarship with further in-depth studies of oil upstream policy choices in producer countries worldwide. This should develop our empirical knowledge further and also relativize the one-size-fits-all solution of economic liberalization, which has been heretofore widely professed by the Bretton Woods institutions.

Furthermore, if data availability might permit, the revised analytical framework proposed in this book could be tested with better proxies and on a longer time span. In this respect, the cooperation between industry and academia would be highly beneficial for both sides. More transparency and data sharing between oil companies and academic scholars are prone to bring about better policy designs with positive upshots in the long run and also be conducive to more reliable data analyses embedded in a conceptual and/or theoretical frame.

The analytical framework and thereby the research for this book were tailored to oil upstream sector policies. Yet this framework may also be applicable to gas. For such amendments, the differences between the oil and gas upstream have to be recognized and the impact of these discrepancies on the mechanisms triggering diverse policy choices need to be examined.

These are but a few distinguishable research avenues to be further explored. The literature on the determinants of oil upstream policies is still in its infant stages, while the prominence of the topic is apparent. In the pursuit of an energy security agenda, the international community is to dedicate increasingly more attention to the oil hotspots and their oil upstream policies in the near future. Nolens-volens the research community will soon do the same.

6.3.2 Quo Vadis?

The topic of this book is timely. Just in fall 2012, the International Energy Agency released a report about Iraq, which is emerging as the third largest exporter of oil worldwide. With plans to increase oil production vastly, the country is now forging its control structures in the oil sector. It is fundamental that it selects the upstream sector policies which are best suited given the conditions on the ground. The way the oil production sector kicks off is decisive not only for the oil upstream industry of Iraq in the medium to long run, but also for the economic and societal future of the country.

Based on the revised analytical framework proposed in this book, international organizations and industry professionals—vying for liberalization —should better grasp why and when oil-rich countries are likely to open up to foreign investors or rather keep their upstream closed. They should also become more open to alternative policies such as mixed or closed upstream sector policies. In this respect, this research has shown that state oil companies are not necessarily inefficient and sluggish organizations anymore—an argument which has been made by other scholars and prestigious magazines before, yet still with little appeal to the larger public. The case studies in this book also revealed that both a closed and an open oil upstream sector can be run efficiently as long as the policy-makers consider the contextual parameters in place and tend to make economically rational decisions.

As long as producer countries are willing and able to sustain an efficient and profit-oriented NOC, their interests—both of commercial and, if need be, of non-commercial nature (see section 2.1.2)—may be better observed than through foreign operators in the oil upstream. The government retains control over production and is able to get the most of the national oil reserves (hopefully) for the benefit of the people in the respective country, while offering operational freedom to the NOC. If, however, the national operator is not at the stage of producing by itself in consideration of the geological conditions in place and the technological requirements, oil producing countries may wish to open their upstream sector to foreign investors and make their NOC collaborate with IOCs on the ground, e.g. in the form of a production-sharing agreement. Alternatively, producer states may devolve (some of) the more complex operations to the IOCs or risk service operators altogether under the legal form of service contracts. Such partnerships between NOCs and IOCs are fruitful for both sides: the former may and should grow their technical capabilities and capitalize on the know-how at stake, whereas the latter are able to use their assets towards company profit.

This book should be relevant to countries assessing their experience with existing control structures as well as the many countries in the process of joining the "petroleum club" of oil producing nations—e.g. Sierra Leone, Sao Tome e Principe, the Democratic Republic of Congo or Madagascar on the African continent, Timor-Leste or Papua New Guinea in Asia, to name just a few. The control structures forged in the oil upstream industry in the present will have serious implications for the economic and societal evolution (or involution) of these countries in the future, as well as for diverse policy areas such as security and economic development, and environment.

This book should serve as a guide for policy-makers, industry professionals and/or international consultants as to the factors which need to be considered and optimized when designing a sustainable policy for the oil upstream industry of individual producer countries around the world. As for those liberalization promoters, this research may be an eye-opener as to when alternative policy choices such as a mixed or a closed oil upstream sector policy are feasible and preferred by governments of producer countries, and also, under which constellation of factors foreign investors may be allowed to step in and relish control rights.

References

Blair, John, *The Control of Oil* (Pantheon Books, New York, 1976).

International Energy Agency, *Iraq Energy Outlook*, October 9, 2012, available at www.worldenergyoutlook.org/iraq/(status: October 21, 2013).

Annex

Annex 1 NOCs abroad

	NOC	Country	Overseas since ...
1	YPF	Argentina	2004
2	OMV	Austria	1985
3	Petrobras	Brazil	1972
4	Petro-Canada	Canada	1979
5	CNPC	China	1994
6	Sinopec	China	2001
7	CNOOC	China	2009
8	ONGC	India	1989
9	IndianOil	India	2002
10	Eni	Italy	1955
11	Petronas	Malaysia	1990
12	Statoil	Norway	1992
13	Norsk Hydro	Norway	1999
14	Lukoil	Russia	1994
15	Gazprom	Russia	2002
16	Yukos	Russia	2001
17	Tatneft	Russia	1989
18	Repsol	Spain	1987
19	CPC	Taiwan	2010
20	TPAO	Turkey	1994

Data Sources (Status: June 4, 2012):

- Petrobras (Brazil): www.petrobras.com/en/about-us/global-presence/;
- Petro-Canada (Canada): www.fundinguniverse.com/company-histories/PETROCANADA-LIMITED-company-History.html;
- CNPC (China): www.cnpc.com.cn/en/cnpcworldwide/peru/Peru.htm and www.cnpc.com.cn/en/cnpcworldwide/;
- Sinopec (China): www.sipc.cn/english/s1/;
- Petrochina (China): www.petrochina.com.cn/Ptr/About_PetroChina/Contact_Us/picl.htm;
- CNOOC (China): www.cnoocltd.com/encnoocltd/AboutUs/zygzq/Overseas/and http://sec.gov/Archives/edgar/data/1095595/000095010311001599/dp21925_20f.htm;
- ONGC (India): www.ongcvidesh.com/company.aspx, www.indianoilandgas.com/data-pdfs/PSU-OI.pdf and www.gasandoil.com/news/2010/01/cns100185;
- IndianOil (India): www.defenddemocracy.org/indian-oil-corporation-ioc/and www.iocl.com/aboutus/e_and_p.aspx;

- Pertamina (Indonesia): www.bakerinstitute.org/programs/energy-forum/publications/energy-studies/docs/NOCs/Papers/NOC_Pertamina_Hertzmark.pdf;
- Eni (Italy): www.eni.com/en_IT/company/history/our-history.page;
- KPC (Kuwait): www.kufpec.com/AboutKUFPEC/Pages/History.aspx#myAnchor;
- Petronas (Malaysia): http://news.google.com/newspapers?nid=1309&dat=19970226&id=Mw5PAAAAIBAJ&sjid=SxUEAAAAIBAJ&pg=6200,2402633 and http://unmondedebrut.pagesperso-orange.fr/PETROL_PAYS/TOGO/REFRENCE_TOGO/petronastogo.htm;
- Statoil (Norway): www.framtiden.no/view-document/504-statoil-in-nigeria.html and www.fundinguniverse.com/company-histories/Statoil-ASA-company-History.html;
- Norsk Hydro (Norway): www.hydro.com/en/About-Hydro/Our-history/1991–2005/;
- Lukoil (Russia): www.lukoil.com/new/history/1994;
- Gazprom (Russia): http://gazprom-international.com/en#en/operations/country/vietnam?overlay=true;
- Yukos (Russia): http://theyukosaffair.com/Origins.htm;
- Repsol (Spain): www.repsol.com/es_en/corporacion/conocer-repsol/perspectiva_historica/1927-1985.aspx;
- CPC (Taiwan): www.cpc.com.tw/english/content/index.asp?pno=63;
- TPAO (Turkey): www.tpao.gov.tr/tp2/sub_en/sub_content.aspx?id=81 and http://en.trend.az/capital/energy/2001155.html.

Annex 2 Countries in the sample

	Country	Frequency	Percent
1	Algeria	24	5.53
2	Argentina	12	2.76
3	Azerbaijan	10	2.30
4	Bahrain	5	1.15
5	China	1	0.23
6	Colombia	23	5.30
7	Ecuador	11	2.53
8	Egypt	24	5.53
9	India	2	0.46
10	Indonesia	24	5.53
11	Iran	24	5.53
12	Iraq	24	5.53
13	Kazakhstan	5	1.15
14	Kuwait	24	5.53
15	Libya	24	5.53
16	Malaysia	3	0.69
17	Mexico	24	5.53
18	Nigeria	24	5.53
19	Norway	5	1.15
20	Oman	23	5.30
21	Peru	1	0.23
22	Qatar	24	5.53
23	Saudi Arabia	24	5.53
24	Syria	14	3.23
25	Turkey	3	0.69
26	UAE	24	5.53
27	Uzbekistan	4	0.92
28	Venezuela	24	5.53
	Total	434	100.00

Annex 3 Countries by year in the sample

Country	Year																								Total
	1987	1988	1989	1990	1991	1992	1993	1994	1995	1996	1997	1998	1999	2000	2001	2002	2003	2004	2005	2006	2007	2008	2009	2010	
Algeria	1	1	1	1	1	1	1	1	1	1	1	1	1	1	1	1	1	1	1	1	1	1	1	1	24
Argentina	1	1	1	1	1	1	1	1	1	1	1	1	0	0	0	0	0	0	0	0	0	0	0	0	12
Azerbaijan	0	0	0	0	0	0	0	0	0	0	0	0	0	0	1	1	1	1	1	1	1	1	1	1	10
Bahrain	1	1	1	1	1	0	0	0	0	0	0	0	0	0	0	0	0	0	0	0	0	0	0	0	5
China	0	0	0	0	0	0	1	0	0	0	0	0	0	0	0	0	0	0	0	0	0	0	0	0	1
Colombia	1	1	1	1	1	1	1	1	1	1	1	1	1	1	1	1	1	1	1	1	1	1	0	1	23
Ecuador	1	1	1	1	1	1	1	1	1	1	1	0	0	0	0	0	0	0	0	0	0	0	0	0	11
Egypt	1	1	1	1	1	1	1	1	1	1	1	1	1	1	1	1	1	1	1	1	1	1	1	1	24
India	1	1	0	0	0	0	0	0	0	0	0	0	0	0	0	0	0	0	0	0	0	0	0	0	2
Indonesia	1	1	1	1	1	1	1	1	1	1	1	1	1	1	1	1	1	1	1	1	1	1	1	1	24
Iran	1	1	1	1	1	1	1	1	1	1	1	1	1	1	1	1	1	1	1	1	1	1	1	1	24
Iraq	1	1	1	1	1	1	1	1	1	1	1	1	1	1	1	1	1	1	1	1	1	1	1	1	24
Kazakhstan	0	0	0	0	0	0	0	0	0	0	0	0	0	1	1	0	0	0	0	0	0	1	1	1	5
Kuwait	1	1	1	1	1	1	1	1	1	1	1	1	1	1	1	1	1	1	1	1	1	1	1	1	24
Libya	1	1	1	1	1	1	1	1	1	1	1	1	1	1	1	1	1	1	1	1	1	1	1	1	24
Malaysia	1	1	1	0	0	0	0	0	0	0	0	0	0	0	0	0	0	0	0	0	0	0	0	0	3
Mexico	1	1	1	1	1	1	1	1	1	1	1	1	1	1	1	1	1	1	1	1	1	1	1	1	24
Nigeria	1	1	1	1	1	1	1	1	1	1	1	1	1	1	1	1	1	1	1	1	1	1	1	1	24

Country																									Total
Norway	1	1	1	0	0	0	0	0	0	0	0	0	0	0	0	0	0	0	0	0	0	0	0	0	5
Oman	1	1	1	1	1	1	1	1	0	1	1	1	1	1	1	1	1	1	1	1	1	1	1	1	23
Peru	1	0	0	0	0	0	0	0	0	0	0	0	0	0	0	0	0	0	0	0	0	0	0	0	1
Qatar	1	1	1	1	1	1	1	1	1	1	1	1	1	1	1	1	1	1	1	1	1	1	1	1	24
Saudi Arabia	1	1	1	1	1	1	1	1	1	1	1	1	1	1	1	1	1	1	1	1	1	1	1	1	24
Syria	0	0	0	0	0	0	1	1	1	1	1	1	1	1	1	1	1	1	1	1	0	0	0	0	14
Turkey	1	1	0	0	1	0	0	0	1	0	0	0	0	0	0	0	0	0	0	0	0	0	0	0	3
UAE	1	1	1	1	1	1	1	1	1	1	1	1	1	1	1	1	1	1	1	1	1	1	1	1	24
Uzbekistan	0	0	0	0	0	0	0	0	0	0	0	0	0	0	1	1	1	1	0	0	0	0	0	0	4
Venezuela	1	1	1	1	1	1	1	1	1	1	1	1	1	1	1	1	1	1	1	1	1	1	1	1	24

Annex 4 Data operationalization and sources

Variable	Operationalization	Data source
DEPENDENT VARIABLE:		
Upstream sector policy	**De facto control** (by year): the ratio of the oil production volume by the NOC ("output liquids") to the total oil production per country ("oil production—barrels")	"Output liquids": PIW Top 50 (1987–2010) "Oil production—barrels": British Petroleum's Statistical Review of World Energy 2011 to which the condensates from EIA database, "World Crude Oil Production" (1970–2009) are added
EXPLANATORY VARIABLES:		
Geological conditions	**Ratio offshore:** ratio of "offshore" oilfields out of the total number of oilfields in the country	"Offshore" (dummy variable): Oil and Gas Journal's "Historical Worldwide Oil Field Production Survey" (1980–2007)
	Average depth: average of the "depth" of all oilfields in a country	"Depth": Oil and Gas Journal's "Historical Worldwide Oil Field Production Survey" (1980–2007)
Technical capabilities	**GDP per capita:** "Gross domestic product per capita, current prices" expressed in U.S. dollars (NGDPDPC)	IMF World Economic Outlook 2012
	School enrolment, tertiary (thousands): the number of nationals (in thousands) enrolled in tertiary education	"School enrollment, tertiary (% gross)" from World Bank's Education Statistics Database multiplied by "population, total" from World Bank's World Development Indicators
Oil reliance	**Oil wealth**	"Oil rents (% of GDP)": World Bank—World Development Indicators
	Oil export dependence: "value of oil exports (TXGO)"/"gross domestic product, current prices (NGDPD)"	"Value of oil exports (TXGO)" and "gross domestic product, current prices (NGDPD)": IMF World Economic Outlook 2011

Variable	Source
Oil price	Crude price ($ 2010)	BP's Statistical Review of World Energy 2011
Executive constraints	POLCON (political constraints index—"POLCON III")	"POLCON III": Henisz' Polcon Dataset 2010
Corruption	Control of corruption	Political Risk Services—International Country Risk Guide data (1984–2010)
CONTROL VARIABLES:		
International institutions	OPEC (dummy variable)	Self-coded based on www.opec.org (OPEC Member Countries)
Regional diffusion effect	MENA (dummy variable)	Self-coded based on World Bank's definition of MENA
Policy inertia	Lag DV: Lag of dependent variable	Generated in Stata
NOC ownership—Type 1	Own 1: 100% state-owned NOC (dummy variable)	Self-coded based on PIW data (1987–2010)
NOC ownership—Type 2	Own 2: majority state-owned NOC (dummy variable)—baseline	Self-coded based on PIW data (1987–2010)
NOC ownership—Type 3	Own 3: minority state-owned NOC (dummy variable)	Self-coded based on PIW data (1987–2010)

Annex 5 Summary statistics

Variable	Obs.	Mean	Std. Dev.	Min	Max
De facto control	434	0.733664	0.225132	0.11	1
Own 1	434	0.917051	0.276124	0	1
Own 3	434	0.0138249	0.1168984	0	1
MENA	434	0.59447	0.491561	0	1
OPEC	434	0.615207	0.487108	0	1
Average depth (thousands)	433	7.474459	1.814932	3.225	11.146
Ratio offshore	434	0.206936	0.264411	0	1
School enrolment, tertiary (thousands)	273	839,508.4	102,3512	2,720.037	529,6951
GDP per capita (current prices, $US million)	417	0.008599	0.012165	0.000161	0.076435
Oil rents of GDP	401	23.4783	18.55838	0.21	105.96
Oil exports of GDP	412	23.97816	18.25787	0	74
Oil crude prices (2010, $US)	434	41.48313	21.1971	17.01	98.5
Polcon III	378	0.155291	0.189689	0	0.69
PRSCorruption	412	2.418588	0.919621	0	6

Annex 6 Correlation matrix

	1	2	3	4	5	6	7	8	9	10	11	12	13	14
1. De facto control	1													
2. own1	-0.2917	1												
3. own3	-0.0824	-0.2956	1											
4. MENA	0.1894	-0.0769	-0.1018	1										
5. OPEC	-0.2506	0.3727	-0.1102	0.4349	1									
6. Average depth (thousands)	-0.06	0.1833	-0.0599	-0.1825	-0.1165	1								
7. Ratio offshore	-0.2206	0.0571	-0.0775	0.1506	0.0321	0.1138	1							
8. School enrolment (thousands)	-0.0359	0.1772	0.0684	-0.4383	-0.1705	-0.0898	-0.2562	1						
9. GDP per capita ($US million)	-0.0262	-0.0732	-0.0006	0.379	0.2526	0.0356	0.5306	-0.3824	1					
10. Oil rents of GDP	-0.1519	-0.0629	-0.1236	0.5242	0.4073	-0.0017	0.2293	-0.5126	0.2811	1				
11. Oil exports of GDP	-0.1358	-0.0915	-0.115	0.6104	0.4703	-0.1205	0.2417	-0.5764	0.4273	0.8768	1			
12. Oil crude price (2010, $US)	-0.1826	0.0969	-0.1041	-0.0314	-0.0079	0.0329	0.0753	0.2721	0.1003	0.1923	0.1659	1		
13. Polcon III	-0.0801	-0.0167	0.1827	-0.5369	-0.3090	0.2535	-0.2961	0.2952	-0.2295	-0.4556	-0.4774	-0.0401	1	
14. PRS corruption	-0.0964	-0.3150	-0.0012	-0.0678	-0.2567	0.1473	0.0671	-0.1910	0.0941	-0.2035	-0.1876	-0.2115	0.1247	1

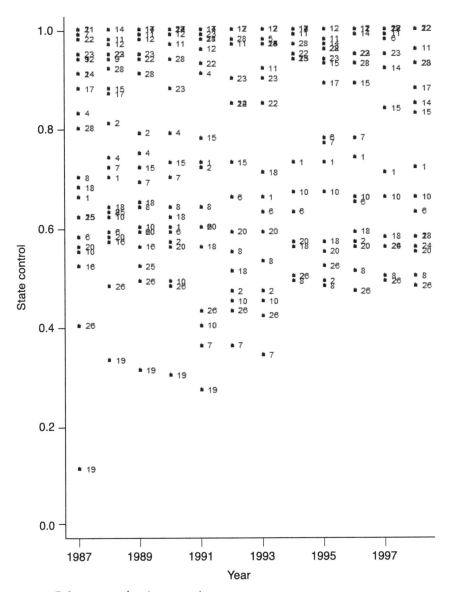

Annex 7 State control ratio versus time

Key: 1 = Algeria; 2 = Argentina; 3 = Azerbaijan; 4 = Bahrain; 5 = China; 6 = Colombia; 7 = Ecuador; 8 = Egypt; 9 = India; 10 = Indonesia; 11 = Iran; 12 = Iraq; 13 = Kazakhstan; 14 = Kuwait; 15 = Libya; 16 = Malaysia; 17 = Mexico; 18 = Nigeria; 19 = Norway; 20 = Oman; 21 = Peru; 22 = Qatar; 23 = Saudi Arabia; 24 = Syria; 25 = Turkey; 26 = UAE; 27 = Uzbekistan; 28 = Venezuela.

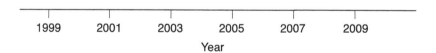

Year

1999   2001   2003   2005   2007   2009

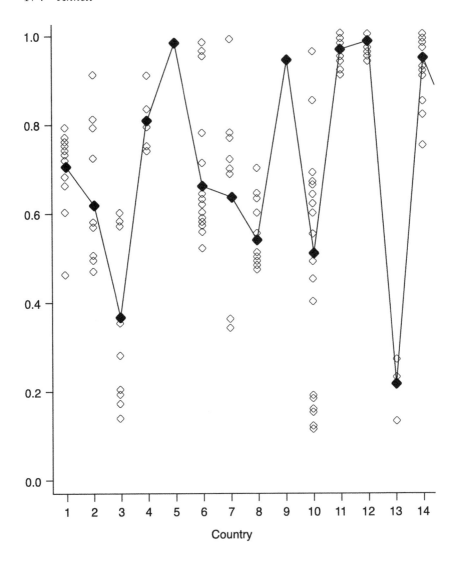

Annex 8 State control ratio—variation across countries

Key: 1 = Algeria; 2 = Argentina; 3 = Azerbaijan; 4 = Bahrain; 5 = China; 6 = Colombia; 7 = Ecuador; 8 = Egypt; 9 = India; 10 = Indonesia; 11 = Iran; 12 = Iraq; 13 = Kazakhstan; 14 = Kuwait; 15 = Libya; 16 = Malaysia; 17 = Mexico; 18 = Nigeria; 19 = Norway; 20 = Oman; 21 = Peru; 22 = Qatar; 23 = Saudi Arabia; 24 = Syria; 25 = Turkey; 26 = UAE; 27 = Uzbekistan; 28 = Venezuela.

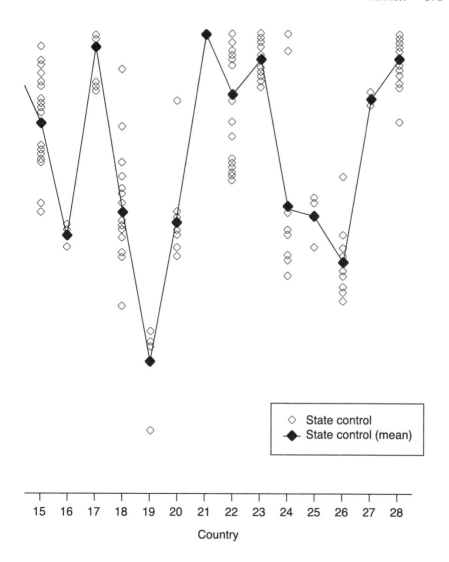

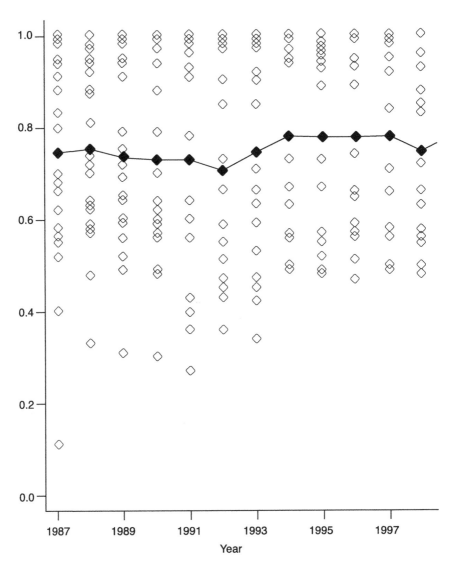

Annex 9 State control ratio (mean) versus time

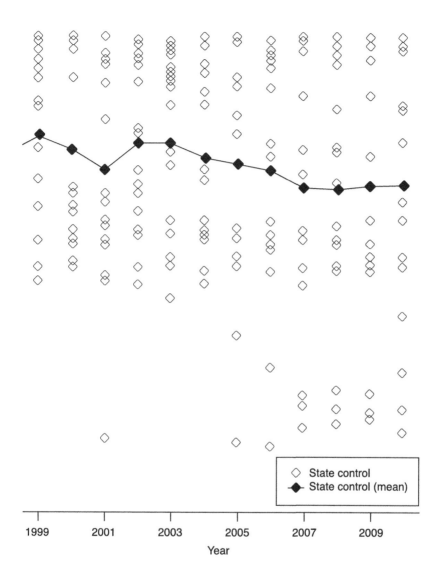

Annex 10 Regression models 1–7

Variables	(1) De facto state control	(2) De facto state control	(3) De facto state control	(4) De facto state control	(5) De facto state control	(6) De facto state control	(7) De facto state control
De facto state control lag	0.331***	0.317***	0.376***	0.386***	0.331***	0.356***	0.355***
	(0.0392)	(0.0397)	(0.0482)	(0.0468)	(0.0401)	(0.0556)	(0.0542)
own1	0.0783***	-0.0587**	0.0171	0.00137	0.0788***	-0.0314	-0.00666
	(0.0241)	(0.0276)	(0.0294)	(0.0339)	(0.0247)	(0.0373)	(0.0350)
own3	-0.0349	-0.130**	-0.157*	-0.168*	-0.0455	-0.239**	-0.178
	(0.0477)	(0.0525)	(0.0899)	(0.0912)	(0.0492)	(0.0935)	(0.111)
OPEC	0.0906***	0.149***	0.0781***	-0.0543***	0.170***	0.0145	-0.0261
	(0.0133)	(0.0176)	(0.0187)	(0.0206)	(0.0185)	(0.0197)	(0.0211)
Ratio offshore	-0.168***	-0.164***				-0.272***	-0.247***
	(0.0239)	(0.0255)				(0.0511)	(0.0470)
Average depth (thousand meters)	-0.0273***	-0.00750**				-0.00346	-0.00246
	(0.00354)	(0.00344)				(0.00457)	(0.00471)
School enrolment, tertiary (thousands)			-1.35e-08*	-2.02e-05**		-9.68e-06	-2.31e-05**
			(8.06e-06)	(8.50e-06)		(1.01e-05)	(1.11e-05)
GDP per capita ($US million)					0.978**		
					(0.496)		
Oil rents (of GDP)							
Oil exports (of GDP)							
Oil crude price (2010 $US)							

	(1)	(2)	(3)	(4)	(5)	(6)	(7)
POLCON III							
POLCON III * Oil rents (of GDP)							
POLCON III * Oil exports (of GDP)							
PRS corruption							
MENA		0.0131		0.128***			0.0757***
		(0.0132)		(0.0222)			(0.0212)
Constant	0.621***	0.592***	0.456***	0.473***	0.329***	0.586***	0.514***
	(0.0438)	(0.0466)	(0.0397)	(0.0500)	(0.0289)	(0.0597)	(0.0607)
Observations	432	432	272	272	416	271	271
Number of countries	27	27	27	27	28	26	26
Log likelihood	245.66876	246.75291	136.75489	138.63762	225.04158	142.74647	144.91577

Standard errors in parentheses *** p < 0.01, ** p < 0.05, * p < 0.1

Annex 11 Regression models 8–14

Variables	(8) De facto state control	(9) De facto state control	(10) De facto state control	(11) De facto state control	(12) De facto state control	(13) De facto state control	(14) De facto state control
De facto state control lag	0.301***	0.317***	0.321***	0.295***	0.257***	0.260***	0.296***
	(0.0407)	(0.0409)	(0.0398)	(0.0411)	(0.0431)	(0.0430)	(0.0422)
own1	-0.0192	0.00446	0.0217	-0.0262	0.0340	0.0455*	-0.0279
	(0.0239)	(0.0256)	(0.0251)	(0.0265)	(0.0260)	(0.0271)	(0.0313)
own3	-0.123**	-0.0754	-0.0385	-0.165***	-0.0984*	-0.0949*	-0.179***
	(0.0494)	(0.0506)	(0.0502)	(0.0515)	(0.0523)	(0.0527)	(0.0588)
OPEC	0.120***	0.132***	0.0865***	0.196***	0.0687***	0.0656***	0.0912***
	(0.0156)	(0.0156)	(0.0146)	(0.0206)	(0.0152)	(0.0151)	(0.0250)
Ratio offshore							
Average depth (thousand meters)	-0.0124***	-0.0101***					
	(0.00350)	(0.00340)					
School enrolment, tertiary (thousands)							
GDP per capita ($US million)	-3.147***	-1.380***					
	(0.541)	(0.520)					
Oil rents (of GDP)			0.000647*				
			(0.000364)				

	(1)	(2)	(3)	(4)	(5)	(6)	(7)
Oil exports (of GDP)				-9.44e-05			
				(0.000419)			
Oil crude price (2010)			-0.000843***	-0.000681**			
			(0.000295)	(0.000282)			
POLCON III					0.0837**	0.0873**	0.272***
					(0.0351)	(0.0357)	(0.0700)
POLCON III * Oil rents (of GDP)							-0.00933***
							(0.00255)
POLCON III * Oil exports (of GDP)							
PRS corruption						0.00789	
						(0.00742)	
MENA		0.0979***					
		(0.0147)					
Constant	0.547***	0.440***	0.489***	0.495***	0.512***	0.346***	0.477***
	(0.0414)	(0.0409)	(0.0370)	(0.0385)	(0.0355)	(0.0382)	(0.0346)
Observations	415	415	400	411	377	376	364
Number of countries	27	27	28	27	28	27	28
Log likelihood	233.63609	238.11438	227.35458	227.68035	201.5395	198.72268	199.31099

Standard errors in parentheses *** p < 0.01, ** p < 0.05, * p < 0.1

Annex 12 Regression models 15–21

Variables	(15) De facto state control	(16) De facto state control	(17) De facto state control	(18) De facto state control	(19) De facto state control	(20) De facto state control	(21) De facto state control
De facto state control lag	0.278***	0.257***	0.287***	0.233***	0.208***	0.243***	0.229***
	(0.0434)	(0.0440)	(0.0421)	(0.0516)	(0.0564)	(0.0511)	(0.0516)
own1	−0.0180	−0.0324	−0.0506*	−0.0670*	−0.0301	0.0348	0.0280
	(0.0285)	(0.0346)	(0.0297)	(0.0395)	(0.0479)	(0.0346)	(0.0356)
own3	−0.131**	−0.157**	−0.196***	−0.389***	−0.366***	−0.321***	−0.331***
	(0.0566)	(0.0611)	(0.0579)	(0.0947)	(0.0987)	(0.0903)	(0.0904)
OPEC	0.124***	0.141***	0.148***	0.0170	0.0659**	−0.0180	−0.00685
	(0.0203)	(0.0295)	(0.0229)	(0.0220)	(0.0301)	(0.0212)	(0.0225)
Ratio offshore				−0.202***	−0.181***		
				(0.0321)	(0.0362)		
Average depth (thousand meters)				−0.00430	−0.00277	−0.0270***	−0.0268***
				(0.00500)	(0.00491)	(0.00502)	(0.00503)
School enrolment, tertiary (thousands)				2.02e−08*	2.07e−08	9.18e−09	6.43e−09
				(1.15e−08)	(1.39e−08)	(1.04e−08)	(1.07e−08)
GDP per capita ($US million)							
Oil rents (of GDP)				0.00131*			
				(0.000719)			

	(1)	(2)	(3)	(4)	(5)	(6)	(7)
Oil exports (of GDP)					-0.0211 (0.129)		
Oil crude price (2010)				-0.00252*** (0.000600)	-0.00233*** (0.000659)	-0.00213*** (0.000562)	-0.00211*** (0.000570)
POLCON III	0.155*** (0.0599)	0.205*** (0.0788)	0.305*** (0.0617)	0.155** (0.0751)	0.223*** (0.0785)	0.229*** (0.0710)	0.245*** (0.0713)
POLCON III * Oil rents (of GDP)	-0.00363* (0.00220)	-0.00496** (0.00248)	-0.0124*** (0.00226)	-0.00840*** (0.00326)		-0.00777*** (0.00297)	
POLCON III * Oil exports (of GDP)					-0.00443 (0.00319)		-0.00793*** (0.00272)
PRS corruption	0.00253 (0.00704)		0.00631 (0.00736)	-0.00623 (0.0101)	-0.00776 (0.0111)	-0.00716 (0.00922)	-0.0106 (0.00964)
MENA							
Constant	0.499*** (0.0440)	0.516*** (0.0371)	0.472*** (0.0428)	0.739*** (0.0748)	0.671*** (0.0963)	0.769*** (0.0702)	0.785*** (0.0704)
Observations	363	358	358	239	233	239	233
Number of countries	27	27	27	25	24	25	24
Log likelihood	201.35701	200.33786	200.3457	133.27344	129.65268	131.20723	128.57069

Standard errors in parentheses *** $p < 0.01$, ** $p < 0.05$, * $p < 0.1$

Annex 13 Regression models 22–25

Variables	(22) De facto state control	(23) De facto state control	(24) De facto state control	(25) De facto state control
De facto state control lag	0.223***	0.233***	0.225***	0.257***
	(0.0448)	(0.0543)	(0.0442)	(0.0436)
own1	-0.0261	0.0594*	0.0603**	0.0416
	(0.0290)	(0.0335)	(0.0253)	(0.0292)
own3	-0.166***	-0.252**	-0.138***	-0.158***
	(0.0550)	(0.126)	(0.0535)	(0.0578)
OPEC	0.0838***			
	(0.0178)			
Ratio offshore		-0.200***		
		(0.0349)		
Average depth (thousand meters)	-0.0134***	-0.00545	-0.00613*	-0.0194***
	(0.00399)	(0.00468)	(0.00359)	(0.00408)
School enrolment, tertiary (thousands)		-8.04e-09		
		(1.15e-08)		
GDP per capita (million U.S. dollars)	-2.263***		0.413	
	(0.638)		(0.647)	
Oil rents (of GDP)				

	(1)	(2)	(3)	(4)
Oil exports (of GDP)	0.00107* (0.000643)			
Oil crude price (2010)	-0.00157*** (0.000476)	-0.00103** (0.000413)	-0.00170*** (0.000569)	-0.000520 (0.000416)
POLCON III	0.222*** (0.0627)	0.199*** (0.0529)	0.292*** (0.0708)	0.233*** (0.0560)
POLCON III * Oil rents (of GDP)		-0.00632*** (0.00201)	-0.00806*** (0.00270)	-0.00498** (0.00213)
POLCON III * Oil exports (of GDP)	-0.00617*** (0.00222)			
PRS corruption	-0.00669 (0.0110)	-0.00815 (0.00733)	-0.00631 (0.0100)	0.00808 (0.00743)
MENA		0.155*** (0.0169)	0.110*** (0.0207)	
Constant	0.644*** (0.0668)	0.460*** (0.0521)	0.663*** (0.0755)	0.606*** (0.0552)
Observations	357	353	239	353
Number of countries	26	26	25	26
Log likelihood	203.83234	203.71954	133.94786	205.17296

Standard errors in parentheses *** $p < 0.01$, ** $p < 0.05$, * $p < 0.1$

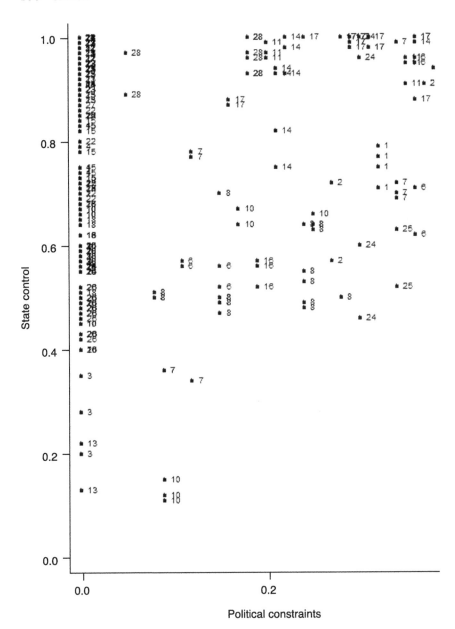

Annex 14 Political constraints versus state control ratio

Key: 1= Algeria; 2 = Argentina; 3 = Azerbaijan; 4 = Bahrain; 5 = China; 6 = Colombia; 7 = Ecuador; 8 = Egypt; 9 = India; 10 = Indonesia; 11 = Iran; 12 = Iraq; 13 = Kazakhstan; 14 = Kuwait; 15 = Libya; 16 = Malaysia; 17 = Mexico; 18 = Nigeria; 19 = Norway; 20 = Oman; 21 = Peru; 22 = Qatar; 23 = Saudi Arabia; 24 = Syria; 25 = Turkey; 26 = UAE; 27 = Uzbekistan; 28 = Venezuela.

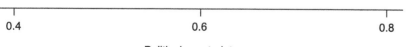

0.4   0.6   0.8

Political constraints

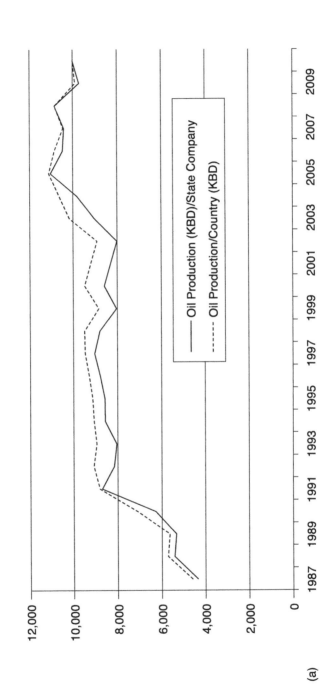

(a)

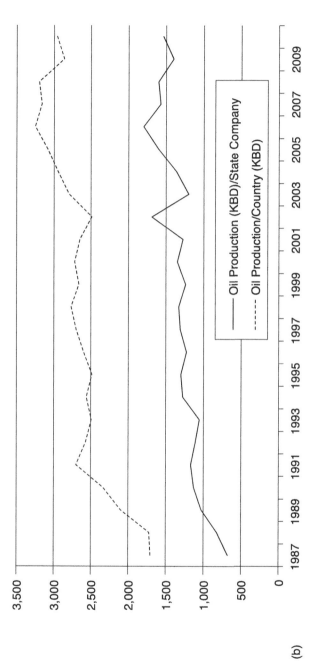

Annex 15 Control structures in (a) Saudi Arabia versus (b) Abu Dhabi

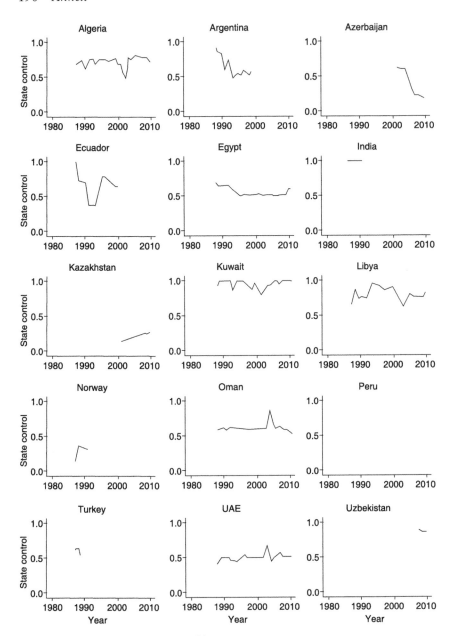

Annex 16 Time versus state control by country

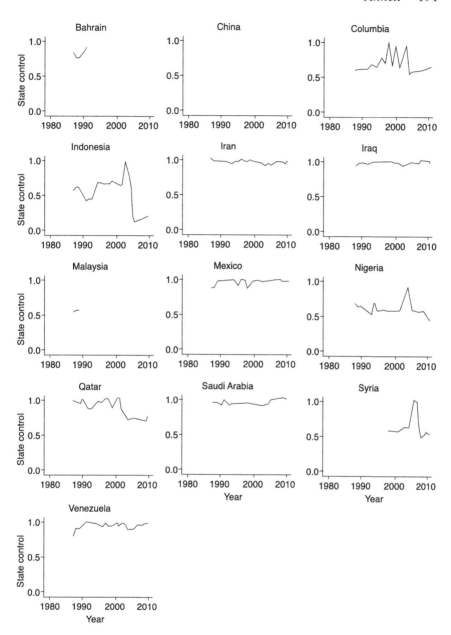

Annex 17 Questionnaire on the oil sector organization and upstream strategies in Saudi Arabia and the United Arab Emirates

Interviewee:

Affiliation:

Date:

Questionnaire on the oil sector organization and upstream strategies in Saudi Arabia and the United Arab Emirates

This interview looks into the institutional structures of the petroleum sector in Saudi Arabia and the United Arab Emirates (UAE) since oil nationalizations in the 1960s–1970s. The main interest lies in the upstream sector—more specifically, the conditions under which foreign oil companies have operated in the exploration and production sector, and the motivations behind host country's preferred strategy in the upstream.

This interview is structured in four parts. In the *first* part, the aim is to map the organization of the petroleum sector, identify the main actors and determine the role of the state and the National Oil Company (NOC) in the petroleum industry. The *second* part looks at the role of International Oil Companies (IOCs) while addressing their mandate in the oil upstream sector of the respective country, and understanding the requirements for control rights. The *third* section focuses on the benefits and (potential) pitfalls of the oil upstream strategy pursued in the respective country. The *fourth* part deals with the relevance of the petroleum industry to the national economy and finally, the potential for sustainable development.

All information provided in this interview is treated as confidential and only used for academic purposes.

Part 1: Oil sector organization

Q.1.1. In your opinion, what are the main institutions in the petroleum sector in:
Saudi Arabia?
a) Ministry of Petroleum and Mineral Resources
b) Supreme Council for Petroleum and Mineral Affairs
c) NOC (Saudi Aramco)
d) IOCs
e) Other: . . .

The UAE?

a) The Ministry of Energy
b) The Supreme Petroleum Council
c) NOCs (ADNOC: Abu Dhabi National Oil Company; ENOC: Emirates National Oil Company)
d) IOCs
e) Other: . . .

Q.1.2. What role does the state play in the petroleum sector in Saudi Arabia/the UAE?
Follow-up Q.1.2: Who is the state in this case?
a) The political leader
b) The government

    c) Members of the royal family
    d) The elites
    e) Other: . . .

Please expand on this.

Q.1.3. What function does the NOC have in Saudi Arabia/the UAE?
- In the petroleum sector?
- At the national level?

Q.1.4. Which institution decides over the clauses and partners of an upstream contract in Saudi Arabia/the UAE?

Q.1.5. To what extent do (patrimonial) networks have a role to play in the petroleum sector?

0 = Not at all	3 = To a moderate extent
1 = To a small extent	4 = To a great extent
2 = To some extent	5 = To a very great extent

Please expand on this.

Q.1.6. Are "brokers" (intermediaries) involved in the petroleum industry?

0 = Not at all	3 = To a moderate extent
1 = To a small extent	4 = To a great extent
2 = To some extent	5 = To a very great extent

Part 2: The role of foreign oil companies in the upstream sector

Q.2.1. Which IOCs have been operating in the upstream sector in Saudi Arabia/ the UAE?

Q.2.2. Which (of the abovementioned) IOCs have had equity shares in the upstream consortium?
Follow-up Q.2.2: What is the percentage of control in the upstream consortium?

Q.2.3. What are the requirements for an IOC in order to qualify for a production-sharing agreement in Saudi Arabia/the UAE?

Q.2.4. What is the upper threshold for upstream control (percentage) which can be granted to an IOC in Saudi Arabia/the UAE?
Follow-up Q.2.4.: Legal vs. in practice?

Q.2.5. What are the challenges of working with IOCs in an oil production consortium in Saudi Arabia/the UAE?
Follow-up Q.2.5.: What are the benefits?

Q.2.6. Under what conditions is an IOC preferred to an NOC for an upstream contract?

Q.2.7. What are the relevant criteria speaking in favor of an IOC to the detriment of other IOCs for an upstream contract?
    a) Technological know-how
    b) Domestic economic interests
       Follow up Q.2.7.: If so, of what sort?

    c) Domestic political interests
       Follow up Q.2.7.: If so, of what sort?
    d) International influences
       Follow up Q.2.7.: If so, of what sort?
    e) Other: . . .

Q.2.8. What leverage does OPEC have in the decision-making process over the upstream strategy in Saudi Arabia/the UAE?
Follow-up Q.2.8.: How about other regional/international organizations?

Q.2.9. How does the technological know-how of the NOC compare with that of the main IOC operating in the upstream in Saudi Arabia/the UAE?

Q.2.10. In your opinion, what is the optimal foreign investor in the oil upstream in Saudi Arabia/the UAE?

Part 3: The oil exploration and production strategy

Q.3.1. Why has the oil production strategy been upheld for the past 25 years?

Q.3.2. What are the core advantages of the current oil upstream strategy in Saudi Arabia/the UAE by comparison with that of the UAE/Saudi Arabia?
Follow-up Q.3.2.: What about drawbacks?

Q.3.3. Are there any domestic actors or groups (in the society, political apparatus, economic arena etc.) opposing the current oil production strategy?
Follow-up Q.3.3.: If so, why?

Part 4: The relevance of the oil sector (present and future)

Q.4.1. How are the revenues stemming from the petroleum sector distributed nationally?
Follow-up Q.4.1.: Are there any investments abroad/sovereign wealth funds?

Q.4.2. What is the contribution of the upstream sector to the state's wealth?
Follow-up Q.4.2.: What is the value of oil to the people?

Q.4.3. How do you assess the position of Saudi Arabia/the UAE in the international market by comparison with other oil producer states?

Q.4.4. What is the potential of Saudi Arabia/the UAE to diversify away from the hydrocarbon sector?
Follow-up Q.4.4.: If existent, in what direction?

Index